Identification Guide to

BIRDS of BRITAIN and NORTHERN EUROPE

PETER GOODFELLOW

This edition published in the United Kingdom in 2021 by John Beaufoy Publishing,
11 Blenheim Court, 316 Woodstock Road, Oxford OX2 7NS, England
www.johnbeaufoy.com

10 9 8 7 6 5 4 3 2 1

ISBN 978-1-912081-98-1

PICTURE CREDITS

Rudi Debruyne 100(lower right). Sławek Staszczuk (photoss@hotmail.co.uk) 133(inset). All other photographs by Paul Sterry except for those on the following pages, which were supplied by Nature Photographers Ltd:
Frank Blackburn 90(right). Laurie Campbell 10(upper). Colin Carver 74(right). Ernie Janes 71(upper), 80(right). Andrew Merrick 100(upper right). Hugh Miles 70(upper). Philip Newman 69(lower), 114(upper). David Osborn 18(lower), 46(upper), 48(lower), 78(upper & lower). WS Paton 71(lower). Richard Revels 13(upper), 61(lower), 63(upper & lower), 66(upper), 67(upper), 68(left). Don Smith 84(left), 86(upper). E Thompson 64(lower), 84(right). Roger Tidman 39(upper), 56(upper), 79(upper), 81(lower), 89(lower), 93(right), 108(left), 142(right), 150(upper).

Maps prepared by the author and created by Martin Brown.

Edited, designed and typeset by D & N Publishing, Baydon, Wiltshire, UK

Printed and bound in Malaysia by Times Offset (M) Sdn. Bhd.

CONTENTS

INTRODUCTION

If you see a flash of colourful plumage, hear a strange call or witness a striking form of behaviour, you may well stop and say, 'I wish I knew what that bird was!' It is hoped that this book will solve that problem, acting as an introductory guide to help residents, travellers and visitors across the region identify the birds they are most likely to see and hear, at any time of the year. In all, about 150 species have been included, which represent a wide variety of breeding species, plus others that visit these shores on migration in spring and autumn.

For every species covered, the English common name and scientific name are provided, followed by a paragraph on each of description, habits and distribution, a distribution map and one or two photographs. A 'Fact File' lists the species' length, wingspan, weight, food, number and colour of eggs, habitat and voice. Unless otherwise stated, the photographs show adult birds. Differences in plumage between the sexes, and at various ages and seasons, are explained in the species description. The distribution section names the countries in which the bird may be found, and also states whether the bird is resident, is a summer breeding visitor and if so where it winters elsewhere, or whether it nests outside our region and so is seen only on migration. An observer needs to read the habitat section alongside this, to narrow down the areas within the country concerned where the bird may be seen. The comments provided on the birds' habits are selective – one comment may emphasise feeding habits, while another may mention something special about breeding. The calendar bar above each species' name shows the months in black in which a species is most likely to be seen in the British Isles.

GEOGRAPHY

The area covered by this book is the northwestern part of the Palaearctic, one of the eight zoogeographic regions into which scientists divide the world according to the similarity of the living organisms they support. Roughly speaking, the northwestern Palaearctic embraces all the land within the region northwards from about 45°N and eastwards to the Kola Peninsula at 40°E – i.e. Iceland, the British Isles (United Kingdom and Eire), northern

BIRDS OF PREY AND OWLS 83

JAN | FEB | MAR | APR | MAY | JUN | JUL | AUG | SEP | OCT | NOV | DEC

PEREGRINE

Falco peregrinus

Broad-based, pointed wings and tapering tail. **ADULT MALE** Dark grey-blue upperparts and tail; buffy-white underparts, finely spotted and barred black. **ADULT FEMALE** Darker and more heavily barred. **JUVENILE** Browner above and underparts streaked, not barred.

Nests on cliff ledges and city towers. Hunts by circling high, spotting prey below, then diving at incredible speed for an aerial kill.

Not Iceland; from British Isles eastwards in suitable habitats. Suffered marked declines in the mid-20th century due to poisons in the food chain; has since recovered well in British Isles.

ABOVE: JUVENILE
LEFT: ADULT

ADULT

FACT FILE

LENGTH 36–48cm WINGSPAN 95–110cm WEIGHT 300–580g (female 15% larger) FOOD Birds, especially seabirds and pigeons EGGS 3–4, buff or cream with heavy red-brown markings HABITAT Chiefly open country with coastal or inland cliffs, including quarries VOICE Shrill *kek-kek-kek-kek*

MAP KEY

ALL YEAR (RESIDENT)

SUMMER (BREEDING)

WINTER (NON-BREEDING)

France, Belgium, the Netherlands, Germany, Poland, the Czech and Slovak republics, Denmark, Sweden, Norway, Finland, the Baltic states (Lithuania, Latvia and Estonia), Belarus and just into Russia.

CLIMATE

The region's climate ranges from Arctic, through boreal (sub-Arctic) to temperate. The powerful Gulf Stream carries warm water across the Atlantic from the Gulf of Mexico to western Europe and, together with the prevailing westerly winds, gives this region milder weather than places in Asia and North America at similar latitudes. For example, Glasgow in Scotland has an average temperature of 3°C in January, yet in Moscow, Russia, on about the same latitude, it drops to -10°C. The Isles of Scilly, at around 49°N and nearly 50km west of the southwestern tip of England, rarely experience frost or snow and have a climate so mild and warm that they are famous for supporting Mediterranean plants although they do suffer from autumn's westerly gales, much to the delight of birdwatchers (p. 8)!

HABITATS

The region covered by this book has a landscape that varies from the flat lowlands of East Anglia in England and the Netherlands, to the mountainous parts of Scotland (>1,300m) and Fennoscandia (>2,000m). In the far north, the habitat is tundra, which merges into boreal forest, or taiga, of birch, willow and Alder stretching from here right across Asia.

Much of northwest Europe was deciduous forest in primeval times, but the vast majority of this has since been cleared by Man to create pasture for Cattle and Sheep, and arable land for crops. The destruction of natural habitats has greatly changed – and is still changing – the region's bird life. Isolated areas of ancient forest, moorland, lowland heath, wetland and grassland now often exist only in national parks and nature reserves.

An appreciation of different habitats is important in order to see as many of northern Europe's birds as possible. Some birds, such as Wrens, will live in a wide range of habitats, but Reed Warblers need reedbeds, Jays live in woodland, Red Grouse are restricted to heather moors, and so on. Many waders can be observed only if you visit coastal or estuarine habitats in spring or autumn. Some species, like House Sparrows and other garden birds seen at birdfeeders, live comfortably with Man; others are very wary of Man, such as some birds of prey. Many habitats are also quite different in winter and summer. An estuary in summer, with a few Herring Gulls, a Grey Heron and a pair of Shelducks, may have a winter population of hundreds of gulls, waders and wildfowl, each of several different species; and a wood in spring, full of the song of tits and summer visitors, will be all but silent in winter. The best advice is to visit your local patch regularly; by becoming familiar with its birds and their habitat preferences, you can then head further afield with more confidence.

TAXONOMY AND NOMENCLATURE

Study of the skeletal structure, plumage, voice and behaviour of birds has helped give order to the 10,000 or so species found worldwide. Nowadays, DNA testing is revealing a new understanding of what a species is and where it sits in relation to other birds. For several hundred years, birds and other living organisms have been allocated names derived from Latin or Greek words. Swedish scientist Carl von Linné (or Linnaeus, to use his Latinised name) standardised the naming of species in his book *Systema Naturae*, published in 1735, adopting a neat two-word system instead of the many words some scientists were using. So, the House Sparrow (p. 144) became *Passer* (genus name) *domesticus* (specific name). Later, it was found that an identifiable geographic variation of the species needed the addition of a third name to distinguish it. For example, the Yellow Wagtail of the British Isles became *Motacilla flava flavissima* to identify it clearly (p. 100). Each scientific name is understood as belonging to a single species across the world, no matter what language the observer may speak or what it is commonly called locally. Thus, the scientific name *Passer domesticus* is the same species, whether it is called locally the House Sparrow (English), Moineau domestique (French), Huismus (Dutch), Haussperling (German), Gråsparv (Swedish) or Домовый воробей (Russian).

The sequence in which zoologists place birds is one you will become more familiar with as you use this and other identification guides more. Roughly speaking, it starts with ancient, primitive birds, and ends with the species that have evolved most recently (passerines, i.e. the songbirds). This orderly arrangement is constantly under review, so may differ slightly from one book to another.

BIRD IDENTIFICATION

When attempting to identify a species, there is a great danger that you will focus on one or two things you have seen rather than observing the whole bird. For example, if you spot a red bird in the garden one morning you will not be able to identify it without noting more detail – it could be a Chaffinch, a Bullfinch, a Crossbill, a Linnet or one of several other species. The illustrations below indicate the parts of a bird that are commonly used to describe appearance. Do become familiar with them. A longer, closer look at your garden bird may show that its underparts are red, its forehead, crown and wings are black, its back is grey, and when it flies there is a flash of white at the base of its tail. With this information it can be narrowed down to a single species, as the only bird that fits this description is the Bullfinch (p. 153)!

Very often, a species can be identified reliably only when the observer hears its call or song. In order to aid identification, you will find comments about the sounds many of the birds make in the 'Fact File' and 'Habits' sections of the species descriptions. For example, the leaf warblers (p. 124) can be separated with certainty only when their song is heard (or you have one in the hand!). Try to go out with an experienced bird listener, or obtain a set of recordings of bird voices (*see* 'References and Further Reading', p. 158).

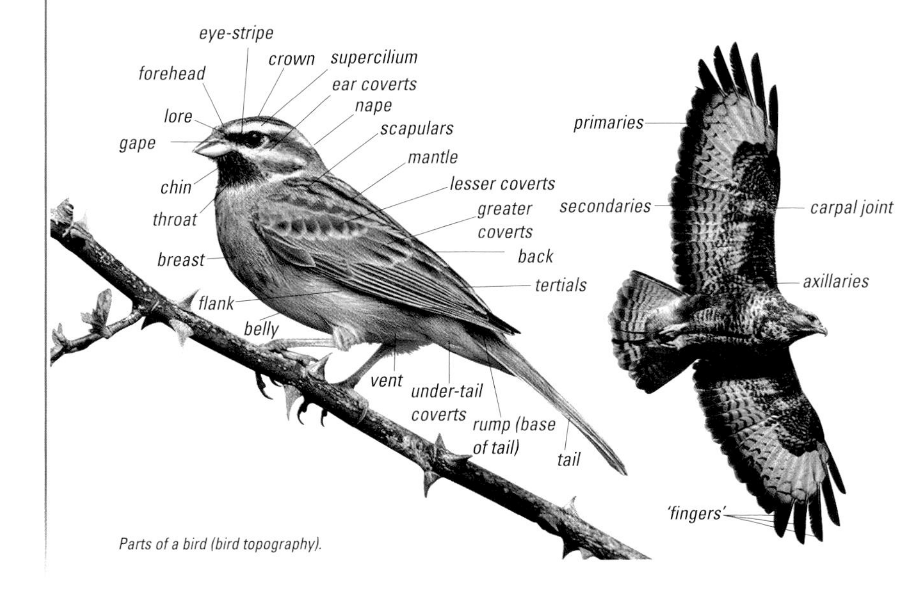

Parts of a bird (bird topography).

GLOSSARY

For a guide to bird topography, *see* the illustration opposite.

Arthropod Any invertebrate with an exoskeleton (the supporting structure covering the outside of the body in many animals), jointed limbs and a segmented body, such as insects, spiders and centipedes.

Bivalves Shellfish with two hinged shells, for example cockles and mussels.

Boreal The northern climatic zone of short summers and long, snowy winters.

Broadleaved Trees that are not coniferous and shed their leaves in the autumn.

Coniferous Trees such as pines, firs, larches, spruces, cedars, junipers and yews, which bear seeds in cones.

Conspecific Belonging to the same species.

Continental shelf The seabed surrounding a continent at depths up to *c.* 220m.

Coppice A dense growth of small trees, regularly trimmed back to stumps to encourage new growth.

Crepuscular Active at twilight or just before dawn.

Deciduous Trees that shed their leaves each year in the autumn.

Defoliating Depriving a plant or tree of its leaves.

Diurnal Active during the day.

Drumming The sound made in flight by the outer tail feathers of a Snipe (p. 54) as it dives in courtship display.

Eyrie The nest of an eagle or other bird of prey.

Fennoscandia The countries of Norway, Sweden and Finland.

Feral Existing in the wild after having been domesticated and then escaped or released.

Gamebird A bird hunted for sport, e.g. grouse, partridges, Pheasant.

Hirundines Birds of the swallow and martin family.

Insectivorous Eats mainly insects, including their larvae and pupae.

Invertebrate Any animal lacking a backbone.

Irruption The sudden entry to a region of large numbers of birds.

Juvenile Young birds that have fledged but have not yet moulted in the autumn into their first adult plumage.

Molluscs Animals including land and sea snails and slugs, and bivalves (shellfish with two shells), such as mussels, cockles and clams.

Nocturnal Active at night.

Omnivorous Eats both plants and animals.

Passage migrant Does not breed in the region but is seen there on its way to breeding or wintering grounds.

Pelagic Lives or occurs on the open sea, not normally coming to land except to breed.

Plankton Small, drifting animals and plants living in the surface layer of the sea.

Raptor A diurnal bird of prey, such as the eagles, hawks and falcons.

Resident Birds that do not migrate to or from the region in which they breed.

Saltmarsh An area of marshy ground that is intermittently inundated with salt water by the tides.

Scandinavia The countries of Denmark, Sweden and Norway.

Saltmarsh is a rich feeding habitat for migrating waders and wildfowl.

Seabird A general term for coastal and pelagic birds such as gulls, terns, shearwaters, petrels, Gannet and Cormorant.

Sedentary Birds that are resident in a certain region and rarely or never move from the territory in which they nest.

Speculum A patch of distinctive, often diagnostic colour on the inner secondaries of dabbling ducks.

***Sphagnum* bog** Any bog with *Sphagnum* mosses as the main vegetation; the vegetation decays to form peat.

Spitzbergen One of the islands of the Norwegian archipelago of Svalbard, sometimes used to describe the whole group.

Steppe Extensive grassy plains, usually without trees.
Summer visitor A bird that migrates (mostly from the south) to breed in the summer, then returns in winter from whence it came.
Taiga Coniferous forest across much of the sub-Arctic, bordered by tundra to the north and steppe to the south.
Tundra A vast treeless zone lying between the Arctic ice cap and the taiga, with a permanently frozen subsoil.
Vagrant A rare, accidental visitor from another region, usually at migration time.
Wader Several closely related families of shore birds, usually with a long bill and legs, which feed in mud at the water's edge or on farmland, e.g. sandpipers, plovers, Curlew.
Winter visitor A bird that migrates (mostly from the north and east) to spend the winter, then returns from whence it came to breed.
Wreck The sudden arrival of exhausted seabirds such as petrels, close to or on land, after prolonged, severe gales at sea.

WHERE TO WATCH BIRDS IN NORTHERN EUROPE

In a region as large as this, there are bound to be places that are particularly good for birdwatching. Birdwatchers are fortunate today in being able to buy field guides to many countries in this region and around the world (*see* 'References and Further Reading', p. 158). There is space here to mention only a few of the most famous sites.

British Isles

- Isles of Scilly, in spring and autumn, especially the latter for rare migrants. The islands are one of the best places to observe migrants from North America, and in September or October it is even possible to see rare visitors from both North America and Asia on the same day!
- Minsmere, Suffolk, one of Britain's finest reserves, owned by the Royal Society for the Protection of Birds (RSPB). Here you can see over 100 species in a day in May.
- Farne Islands, Northumberland, within easy reach by boat on a day trip from Seahouses. It is

Fair Isle, northern Scotland, between the Orkney and Shetland Isles: home of a famous bird observatory since 1948.

LEFT: *Beech woods provide seed (beechmast), which is vital winter food for finches.*
BELOW: *Tundra, the home of Arctic breeding waders and Snowy Owls.*

home to easily observed seabird colonies – a walk among the terns is a must!

France

- Parc Naturel Régional de Brière (Brière Regional Natural Park), near Nantes, one of the best wetlands in Europe, with a park office on Île de Fedrun. The place to see harriers, rails, Black Terns and much more.
- Fôret d'Argonne (Argonne Forest), east of Reims, best known as a splendid place to find up to 10 species of birds of prey.

Germany

- Nationalpark-Wattenmeer (Wadden Sea National Park, a UNESCO site), which takes in all the country's North Sea coast. It is divided into three separate national parks, each with visitor centres.

Poland

- Puszcza Bialowieska (Bialowieska Forest) for rare owls, birds of prey and woodpeckers.
- Biebrzanski Park Narodowy (Biebrza National Park), whose marshes in spring are home to Cranes.

Fennoscandia

- Falsterbo Peninsula, Sweden, long famous as a site for watching the autumn migration, especially of over a dozen species of raptors in September.
- Hardangervidda Nasjonalpark (Hardangervidda National Park), Norway, the largest mountain plateau in northern Europe. The breeding season brings wildfowl, Cranes, waders and buntings.
- Patvinsuon Kansallispuisto (Patvinsuo National Park), Finland, covers thousands of hectares of moorland and ancient forest, and is home to bears, wolves, four species of grouse, waders and owls.

JAN | FEB | MAR | APR | MAY | JUN | JUL | AUG | SEP | OCT | NOV | DEC

BLACK-THROATED DIVER

Gavia arctica

Breeds in Iceland and Scandinavia, plus a few in N Scotland.

SUMMER ADULT

SUMMER ADULT Pale grey head and nape; black throat, white underparts. Striking white patches on back. Wings black. **WINTER ADULT** Upperparts plain dark grey; neck and underparts white. **JUVENILE** As winter adult but with pale feather edges to upperparts.

Swims low. Dives for fish, slipping underwater, not leaping out first like Cormorant (p. 16).

FACT FILE

LENGTH 63–75cm WINGSPAN 100–122cm WEIGHT 1.3kg (winter) to 3.4kg (summer) FOOD Mainly fish EGGS 2; olive or umber, slightly spotted black HABITAT Breeds close to the water's edge on freshwater lakes. Winters on coasts further S VOICE Song: loud, repeated *clooee*

JAN | FEB | MAR | APR | MAY | JUN | JUL | AUG | SEP | OCT | NOV | DEC

RED-THROATED DIVER

Gavia stellata

Breeds in Iceland, N Scotland, Scandinavia. Winters, often in loose flocks, around those coasts and S to Spain and the Black Sea.

SUMMER ADULT

SUMMER ADULT Pale grey head and neck, with rufous-red throat. Upperparts plain, dark grey-brown. **WINTER ADULT AND JUVENILE** Pale grey crown and hind-neck. Face and underparts white. Upperparts finely speckled with white.

Swims low, so in winter often only white on head and neck show. Nests at water's edge.

FACT FILE

LENGTH 55–67cm WINGSPAN 91–110cm WEIGHT 1.0kg (winter) to 1.9kg (summer) FOOD Mainly fish EGGS 2; olive or umber, sparsely spotted blackish HABITAT Breeds on small lakes, tundra pools, forest bogs; fishes on larger lakes or the coast VOICE Silent in winter, but breeding pairs duet with loud calls

JAN | FEB | MAR | APR | MAY | JUN | JUL | AUG | SEP | OCT | NOV | DEC

SLAVONIAN GREBE

Podiceps auritus

SUMMER ADULT

SUMMER ADULT Golden flash on head, reddish neck and body, black wings with small white shoulder and white speculum. **WINTER ADULT AND JUVENILE** Crown down to eye level, hind-neck and back black; wings as summer. Rest of face, neck and underparts white.

Nest attached to aquatic vegetation, often floating. Grey chicks have black- and white-striped heads.

Breeds in Iceland, around Baltic and N Norway, plus a few in N Scotland.

FACT FILE

LENGTH 31–38cm WINGSPAN 46–55cm WEIGHT 300–470g FOOD Mostly insects; also small fish EGGS 3–6; chalky white, becoming stained HABITAT Most breed on shallow lakes with plenty of vegetation. Coastal in winter VOICE Loud, nasal *arrrrrrrr* and trilling courtship duet

JAN | FEB | MAR | APR | MAY | JUN | JUL | AUG | SEP | OCT | NOV | DEC

BLACK-NECKED GREBE

Podiceps nigricollis

SUMMER ADULT

SUMMER ADULT Unmistakable head pattern and colours. Neck and upperparts dull black. Flanks chestnut, belly white. **WINTER ADULT and JUVENILE** Crown down to well below eye level, hind-neck and back black; wings as summer adult, black with white secondaries. Face, neck and underparts white.

Nest is a heap of water weed in shallow water.

Scattered breeding populations in British Isles, S Sweden and rest of region. Winters around W European coasts.

FACT FILE

LENGTH 28–34cm WINGSPAN 56–60cm WEIGHT 220–450g FOOD Insects, molluscs EGGS 3–5; chalky white, becoming heavily stained HABITAT Small, shallow waters with rich vegetation and a good food supply VOICE Repeated *pu-iiii*

JAN | FEB | MAR | APR | MAY | JUN | JUL | AUG | SEP | OCT | NOV | DEC

LITTLE GREBE

Tachybaptus ruficollis

Widespread and mainly resident in British Isles and France, eastwards to S Baltic.

SUMMER ADULT Black cap; chestnut from neck to blackish-brown underparts, fading to paler brown flanks. Wings brown above, white below. **WINTER ADULT** Crown, nape and back dark brown. Wings brown. Rest of plumage visible as it swims is buff. **JUVENILE** Striped head.

Secretive, often discovered only by its loud, whinnying call. When disturbed, will dive or scuttle across water rather than fly. Floating nest is attached to vegetation.

LEFT: *WINTER ADULT*
BELOW: *SUMMER ADULT*

FACT FILE

LENGTH 25–29cm WINGSPAN 40–45cm WEIGHT 130–235g FOOD Insects, larvae, molluscs, small fish EGGS 4–6; chalky white, becoming stained HABITAT Breeds on inland waters, even small ponds or ditches. Some N and E birds winter S and W on larger lakes and the coast VOICE Whinnying call

JAN | FEB | MAR | APR | MAY | JUN | JUL | AUG | SEP | OCT | NOV | DEC

GREAT CRESTED GREBE

Podiceps cristatus

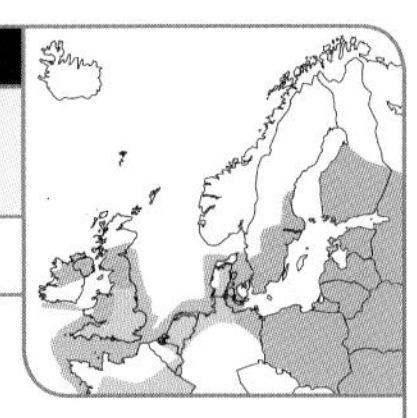

Widespread breeder from British Isles (not N Scotland) to Baltic.

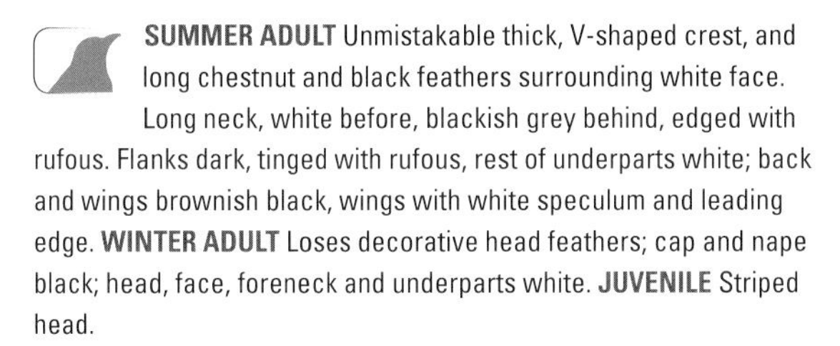

SUMMER ADULT Unmistakable thick, V-shaped crest, and long chestnut and black feathers surrounding white face. Long neck, white before, blackish grey behind, edged with rufous. Flanks dark, tinged with rufous, rest of underparts white; back and wings brownish black, wings with white speculum and leading edge. **WINTER ADULT** Loses decorative head feathers; cap and nape black; head, face, foreneck and underparts white. **JUVENILE** Striped head.

HABITS Erect neck makes it easily distinguishable on open water. Nest is a floating heap of vegetation at the water's edge.

RIGHT: *ADULT AND YOUNG*
BELOW: *ADULT*

FACT FILE

LENGTH 46–51cm WINGSPAN 59–73cm WEIGHT 600–1,400g (male heavier than female) FOOD Mostly fish EGGS 3–5; chalky white, becoming stained HABITAT Breeds and winters on large lakes, reservoirs and gravel pits; also winters on the coast VOICE Far-carrying, rolling *crrrra-ahrr*

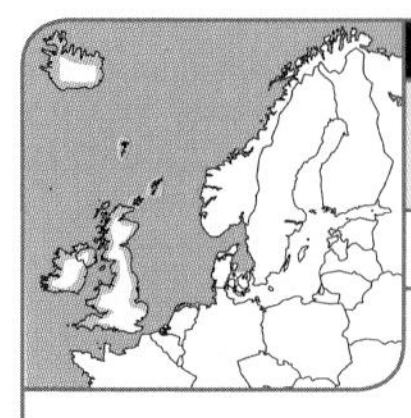

JAN | FEB | MAR | APR | MAY | JUN | JUL | AUG | SEP | OCT | NOV | DEC

FULMAR

Fulmarus glacialis

Colonial breeder on coastal cliffs from the Channel northwards to Iceland and, in a few cases, eastwards to Norway. Huge population growth in past 200 years.

Gull-like, with a white head and body, and grey upperwings, but it is a petrel – note the short yellowish bill with pronounced tubular nostrils.

Flight distinctive, on stiff wingbeats; glides on straight wings, usually low over sea. Pairs for life. One egg laid on bare rock. Chick flies after c. 6 weeks; lives at sea for 4–5 years.

FACT FILE

LENGTH 43–52cm WINGSPAN 101–117cm WEIGHT 600–1,000g FOOD Fish, offal from fishing boats, crustaceans EGGS 1; white, becoming marked with excreta HABITAT Comes to land only to breed. Winters wholly at sea, to c. 45°S VOICE Mates cackle to each other on nesting ledge

JAN	FEB	**MAR**	**APR**	**MAY**	**JUN**	**JUL**	AUG	SEP	OCT	NOV	DEC

MANX SHEARWATER

Puffinus puffinus

A few small colonies in Iceland and Brittany; thousands in Faeroes, and c. 250,000 pairs in colonies around W and N British Isles. All birds migrate SW to overwinter off South American coast.

Uniformly sooty black above and white below; white underwing edged black. Dark on head extends to just below eye.

Glides, or flies on stiff wingbeats, low over the sea, looking black and then white. Dives for food near the surface. Comes to nest only at night to avoid predatory gulls. One fledgling is known to have reached Brazil in 17 days.

FACT FILE

LENGTH 30–35cm WINGSPAN 71–83cm WEIGHT 350–550g FOOD Fish, squid EGGS 1; dull white, becoming stained HABITAT Nests in seaside burrows VOICE Chattering at colony, otherwise silent

JAN | FEB | MAR | APR | MAY | JUN | JUL | AUG | SEP | OCT | NOV | DEC

CORMORANT

Phalacrocorax carbo

Breeds mostly in Iceland, British Isles and N Norway, and (fewer) from France to Poland. British birds disperse locally; more N birds migrate S.

ADULT All black, body glossed blue or green, wings glossed bronze. Summer birds have white on crown and nape. **JUVENILE** Brown with grubby-white underparts; sometimes misidentified as a penguin.

Dives for food with a leap out of the water. Roosts with wings spread to dry. Flies with neck outstretched.

FACT FILE

LENGTH 77–94cm WINGSPAN 121–149cm WEIGHT 2.5–3.5kg FOOD Fish, especially flatfish EGGS 3 or 4; pale blue, patchily covered with chalky white HABITAT Breeds in cliff colonies, and in trees by lakes or coast. Feeds in sandy or muddy estuaries and bays VOICE Deep, guttural calls at colony

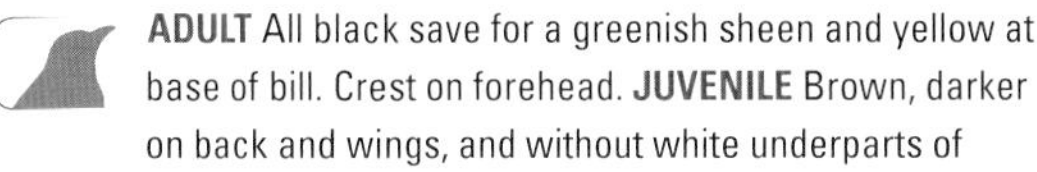

SHAG

Phalacrocorax aristotelis

ADULT All black save for a greenish sheen and yellow at base of bill. Crest on forehead. **JUVENILE** Brown, darker on back and wings, and without white underparts of Cormorant (p. 16). Flight is with quicker wingbeats than Cormorant.

Breeds colonially; nest of twigs and seaweed on a sheltered ledge. Feeds by diving with a leap forward. Roosts singly or in small groups on rocks, often with wings spread.

Resident, with some local coastal dispersal after breeding, in Iceland, British Isles, NW France and Norway.

FACT FILE

LENGTH 68–78cm WINGSPAN 95–110cm WEIGHT 1.8–2.2kg FOOD Fish EGGS 3; pale blue, patchily covered with chalky white HABITAT Coastal, breeding on sea cliffs and mostly feeding over rocky shores VOICE Various grunting sounds at colony

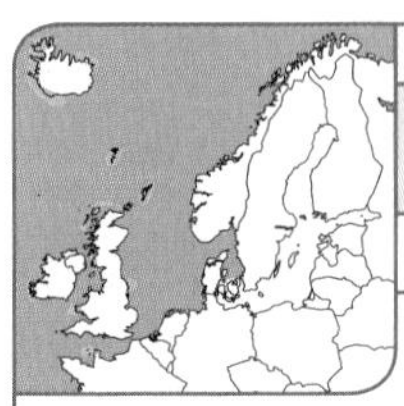

JAN | **FEB** | **MAR** | **APR** | **MAY** | **JUN** | **JUL** | **AUG** | **SEP** | OCT | NOV | DEC

GANNET

Morus bassanus

Breeding colonies in Iceland, Faeroes, Norway and Brittany, but mostly in W and N British Isles (70% of world's total; most of remainder in E Canada). In winter, wanders as far as W Africa.

Region's largest seabird. **ADULT** Mostly pure white but with black wing-tips; yellowish-orange head in breeding season. Long, pointed wings; wedge-shaped tail. **JUVENILE** Entirely grey-brown, covered with fine white speckles; white increases, starting with head and body, until adult plumage is reached after 4 years.

Fishes spectacularly by diving steeply from 10–40m above sea, often in flocks.

FACT FILE

LENGTH 85–97cm WINGSPAN 170–192cm WEIGHT 2.3–3.6kg FOOD Fish EGGS 1; pale blue, covered with chalky white, soon stained HABITAT Pelagic over the N Atlantic continental shelf. Breeds on remote cliff tops VOICE Loud grating sounds at nest

JAN	FEB	**MAR**	**APR**	**MAY**	**JUN**	**JUL**	**AUG**	SEP	OCT	NOV	DEC

WHITE STORK

Ciconia ciconia

Declining. Scattered in E France and Low Countries, more common eastwards to S side of Baltic. Vagrant to British Isles. Bred in UK in 2019 for first time in many years. Much-loved bird of European legend.

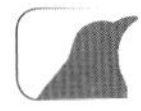

ADULT All-white body and wing coverts; black flight feathers. Red legs and long, pointed red bill. **JUVENILE** Bill has dark tip.

Large stick nest used each year, in trees or on rooftops, towers and pylons. Summer visitor, wintering in tropical Africa. Large migrant flocks gather at E and W Mediterranean's narrows, circling; flocks often seen at plentiful food sources.

FACT FILE

LENGTH 95–110cm WINGSPAN 180–218cm WEIGHT 2.3–4.4kg FOOD Mostly insects and animals EGGS 4, chalky white HABITAT Damp pastureland, marshes, riversides VOICE Rarely heard; loud bill-clappering used in pair's greeting ceremony

JAN | FEB | MAR | APR | MAY | JUN | JUL | AUG | SEP | OCT | NOV | DEC

GREY HERON

Ardea cinerea

Widespread resident in British Isles and France, eastwards to Baltic and coastal Norway; E birds migrate W and S.

ADULT Legs very long, neck long. Back and coverts blue-grey, flight feathers dull black. Head white with black stripe over and behind eye, ending in long nape plumes. Body greyish white with black and white stripes down centre. **JUVENILE** Much greyer, lacking black and white contrast on head; bill greyish horn.

Usually solitary, but a colonial nester in tall trees ('heronries'). In flight, neck is retracted, legs trail beyond tail, and wingbeats are deep and slow.

FACT FILE

LENGTH 84–102cm
WINGSPAN 155–175cm
WEIGHT 1.0–1.8kg
FOOD Mostly fish, eels; also amphibians, small birds, mammals
EGGS 3–5, greenish blue
HABITAT Shallow lake edges, slow rivers, marshes, estuaries
VOICE Loud, harsh *frank*

JAN | FEB | MAR | APR | MAY | JUN | JUL | AUG | SEP | OCT | NOV | DEC

LITTLE EGRET

Egretta garzetta

Neck is curled back when resting and flying; feet project beyond tail in flight. **ADULT** Has thin plumes on breast and back, and 2 long nape feathers. Toes bright yellow. **IMMATURE** Lacks plumes and toes are less bright.

Colonial breeder, often with Grey Herons (p. 20), in trees or bushes. Usually feeds singly, wading and sometimes energetically chasing prey.

Mostly a central and S European breeder. Unusual numbers occurred in Britain in 1989, and many birds stayed; now >1,000 breeding pairs occur here, and c. 5,000 winter on sheltered estuaries. Also increasing in Brittany and the Netherlands.

FACT FILE

LENGTH 55–65cm WINGSPAN 88–106cm WEIGHT 450–550g FOOD Small fish, crustaceans EGGS 3–5, greenish blue HABITAT Breeds by marshes, lakes and lagoons. Winters on sheltered estuaries and coasts VOICE Hoarse *aaaah* when flushed; other harsh calls at colony

Widespread resident, especially in British Isles; less common elsewhere. E birds migrate to moulting areas in W Baltic.

JAN FEB MAR APR MAY JUN JUL AUG SEP OCT NOV DEC

MUTE SWAN

Cygnus olor

Very large swan. **ADULT** All white, with black webbed feet, and orange-red bill with a black base. Knob at bill base is largest on male, which is larger than female. **JUVENILE** Grey-brown, bill dark grey; white plumage in following year.

Most W birds are probably feral from centuries-old introductions. Now even found in towns. Feeds by dipping head or up-ending. Huge nest on ground is protected by aggressive male.

long neck

orange-red bill

pointed tail

ABOVE: *ADULT MALE;* RIGHT: *JUVENILE*

FACT FILE

LENGTH 140–160cm WINGSPAN 200–240cm WEIGHT 8–14kg FOOD Aquatic vegetation EGGS 3–8; white with greyish or blue-green tinge, becoming stained HABITAT Lakes, slow-flowing rivers and canals with shallow water for feeding VOICE Angry hissing; in flight, wings make diagnostic loud, throbbing sound

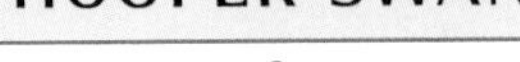

JAN FEB MAR APR MAY JUN JUL AUG SEP OCT NOV DEC

WHOOPER SWAN

Cygnus cygnus

Head usually held erect, not with Mute's curve. **ADULT** All-white plumage, head sometimes stained brown. Square-tipped tail. Basal half of wedge-shaped bill yellow, finishing in a point, the rest black. Legs black. **JUVENILE** Brownish grey with off-white and pink bill.

Pairs breed singly, family staying together in winter flocks. All flight feathers are moulted together, so birds are flightless for 5–6 weeks after breeding. Feeds by up-ending.

Breeds in Iceland and Fennoscandia; c. 75% of the former winter in Britain (mostly Scotland) and Ireland, while the latter winter on North Sea coasts.

FACT FILE

LENGTH 140–160cm WINGSPAN 205–235cm WEIGHT 8–14kg FOOD Aquatic vegetation EGGS 3–5; white or tinged yellow HABITAT Breeds in boggy tundra and upland lakes; winters on lowland farmland VOICE Highly vocal, with bugling calls. Insignificant wing noise, unlike Mute (opposite)

JAN FEB MAR APR MAY JUN JUL AUG SEP OCT NOV DEC

BEWICK'S SWAN

Cygnus columbianus

ADULT All-white plumage relieved only by black legs and yellow and black bill; yellow on bill generally a square or rounded patch, variable in shape, enabling observers to recognise individuals. **JUVENILE** Grey, with pinkish-grey bill.

Winters in flocks. Families keep together for the winter. Adults pair for life.

Winter visitor from Siberian breeding haunts, especially to British Isles (c. 30%) and the Netherlands.

FACT FILE

LENGTH 115–127cm WINGSPAN 170–195cm WEIGHT 3.4–7.2kg FOOD Aquatic plants; also grazes grass and waste potatoes and carrots EGGS 3–5; white or slightly yellow, becoming stained HABITAT Winters on low-lying wet pastures, saltmarshes, lakes VOICE Very vocal, trumpeting and honking

JAN | FEB | MAR | APR | MAY | JUN | JUL | AUG | SEP | OCT | NOV | DEC

GREYLAG GOOSE

Anser anser

Originally bred over most of Europe, but hunting and drainage have reduced wild populations to Scotland, Iceland, Poland, around Baltic and coastal Norway. Reintroduced to England and the Low Countries.

Stocky, with a thick neck and large bill. Back dark grey-brown, paler below, with narrow, darker stripes on flanks and neck. Forewing and underwing coverts noticeably pale grey, very clear in flight. Wide white band at base and tip of tail.

Feral British birds breed near lakes or on their islands, and feed on local grassland. Wild populations nest in marshes and wet tundra sites. Icelandic birds migrate to Britain; Scandinavian and E birds migrate SW as far as Spain. Ancestor of farmyard goose.

FACT FILE

LENGTH 74–84cm WINGSPAN 149–168cm WEIGHT 3–4kg FOOD Grass, fallen grain EGGS 4–6, creamy white HABITAT Farmland, marshes, lakes VOICE Loud cackling

JAN | FEB | MAR | APR | MAY | JUN | JUL | AUG | SEP | OCT | NOV | DEC

CANADA GOOSE

Branta canadensis

Large brown goose, with black head and neck relieved by white blaze from behind eye to meet under chin. Wings dark above and below. White under base of tail. Bill and legs black.

Gander defends breeding territory aggressively, but species is gregarious outside the breeding season, when flocks of hundreds or more form. Can become a pest in large numbers, fouling lakesides. Signals its approach in flight with loud calls.

Native of North America; introduced to Britain in the 17th century, now widespread. Also feral in the Netherlands, Belgium and Scandinavia. Mostly resident, but E birds migrate SW.

FACT FILE

LENGTH 90–100cm WINGSPAN 160–175cm WEIGHT 3.5–5.0kg FOOD Aquatic and waterside vegetation EGGS 5–6, white to creamy white HABITAT Lowland lakes, even in city parks VOICE In flight, deep, loud, trumpeting calls

JAN	FEB	MAR	APR	MAY	JUN	JUL	AUG	SEP	OCT	NOV	DEC

BARNACLE GOOSE

Branta leucopsis

Arctic breeder. Three populations winter here: from Greenland to Ireland and NW Scotland; from Spitzbergen to Solway Firth; from Siberia mostly to the Netherlands.

Distinctive creamy-white face and black cap, neck and breast. Other underparts white, with grey markings on flanks. Back and wing coverts blue-grey, striped black and white. Tail black. Black/white/grey pattern noticeable in flight.

Feeding on farmland has put birds into conflict with farmers. Escapees from wildfowl collections now nest in Britain.

FACT FILE

LENGTH 58–70cm WINGSPAN 120–142cm WEIGHT 1.3–2.4kg FOOD Grazes for grass, roots, seeds EGGS 3–5, creamy white HABITAT Breeds on islands and inaccessible cliffs; winters on coastal grassland and marshes VOICE Barking, yapping; noisy in a flock

JAN	FEB	MAR	APR	MAY	JUN	JUL	AUG	SEP	OCT	NOV	DEC

BRENT GOOSE

Branta bernicla

ADULT DARK-BELLIED

Breeds further N than other geese: 'dark-bellied' (Siberia) winters in the Netherlands, W France and England; 'light-bellied' (Greenland) and Canadian Arctic winters in Ireland.

Small, duck-sized goose. At a distance, looks all black with a large white patch under tail. Two subspecies: 'dark-bellied', with belly almost as dark as back, but mottled; and 'light-bellied', with pale, mottled grey-brown belly.

In winter, feeds on mudflats. Grazes night or day depending on tide. Has moved to graze on grassland.

FACT FILE

LENGTH 55–62cm WINGSPAN 105–117cm WEIGHT 1.2–1.5kg FOOD Grazes grass, underwater vegetation EGGS 3–5, off-white HABITAT Winters along sea coasts and estuaries; breeds on tundra VOICE Monosyllabic *rronk*, becoming a babble in a flock

JAN | FEB | MAR | APR | MAY | JUN | JUL | AUG | SEP | OCT | NOV | DEC

SHELDUCK

Tadorna tadorna

Widespread around region's coasts; less common in Baltic and absent from Iceland.

Large, distinctive duck: white with 'black' head (bottle green in good light), chestnut-orange breast-band, and black flight feathers. Legs pink; bill red, male's larger than female's and with noticeable knob at base.

Food sifted from mudflats by sweeping bill action. To breed it needs a suitable nest-hole, e.g. in a sand-dune, riverbank or Rabbit burrow. After breeding, migrates to traditional moulting areas, especially Heligoland Bight and some British estuaries.

MALES

noticeably red bill of male

chestnut breast-band on white body is diagnostic

orange under-tail coverts

FACT FILE

LENGTH 55–65cm WINGSPAN 100–120cm WEIGHT 562–1,500g FOOD Invertebrates, especially the small snail *Hydrobia* EGGS 8–15, creamy white HABITAT Breeds on sheltered coasts and estuaries VOICE Noisy *ak-ak-ak-ak* in breeding season

JAN | FEB | MAR | APR | MAY | JUN | JUL | AUG | SEP | OCT | NOV | DEC

MALLARD

Anas platyrhynchos

Widespread and common throughout region. Largely resident but retreats from ice.

MALE Very distinctive pattern of green, yellow, white, purple, grey and black; unique central tail feathers; orange legs. After moulting, resembles female with yellowish bill. **FEMALE** Camouflaged brown with darker markings; orange bill marked with black. Striking dark blue speculum with black/white borders.

Ancestor of the farmyard duck. Very adaptable, nesting even in city parks. Like other dabbling ducks, up-ends for food.

LEFT: *FEMALE*
BELOW: *MALE*

FACT FILE

LENGTH 50–60cm WINGSPAN 81–95cm WEIGHT 750–1,450g (males bigger) FOOD Omnivorous EGGS 9–12, greenish buff HABITAT Lakes, ponds, canals, streams, from moorland to coasts VOICE Male, rasping call and low whistle; female, quacks

JAN | FEB | MAR | APR | MAY | JUN | JUL | AUG | SEP | OCT | NOV | DEC

SHOVELER

Anas clypeata

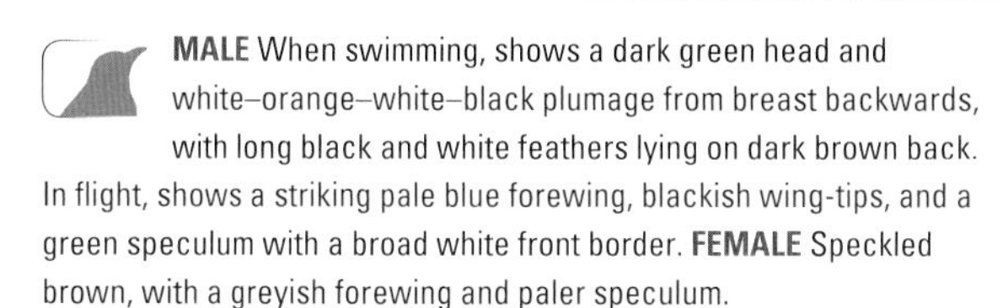

MALE When swimming, shows a dark green head and white–orange–white–black plumage from breast backwards, with long black and white feathers lying on dark brown back. In flight, shows a striking pale blue forewing, blackish wing-tips, and a green speculum with a broad white front border. **FEMALE** Speckled brown, with a greyish forewing and paler speculum.

Feeds by reaching its huge bill forward, filtering tiny food through it as it swims.

Scattered summer breeding populations in Iceland, British Isles and France, eastwards to Baltic and on across Asia. All birds migrate S or W, British birds to S Europe; major wintering areas for E birds are Britain, Ireland and the Netherlands.

TOP: *FEMALE*
ABOVE: *MALE*

FACT FILE

LENGTH 44–52cm WINGSPAN 73–82cm WEIGHT 400–850g FOOD Omnivorous EGGS 9–11, buff or olive HABITAT Lowland marshes and shallow waters VOICE Male, quiet *took*; female, quacks

JAN FEB MAR APR MAY JUN JUL AUG SEP OCT NOV DEC

WIGEON

Anas penelope

Breeds in Arctic or sub-Arctic Iceland, Scotland and Scandinavia eastwards. British and Icelandic birds are resident or make local movements; E birds winter in W and SW Europe.

MALE Unmistakable head colours, pink breast, grey body and black tail end. In flight, shows green speculum and large white patch on forewing. **FEMALE** Rufous-brown body, darker mottled back, rounded head. Both show a noticeable white belly in flight.

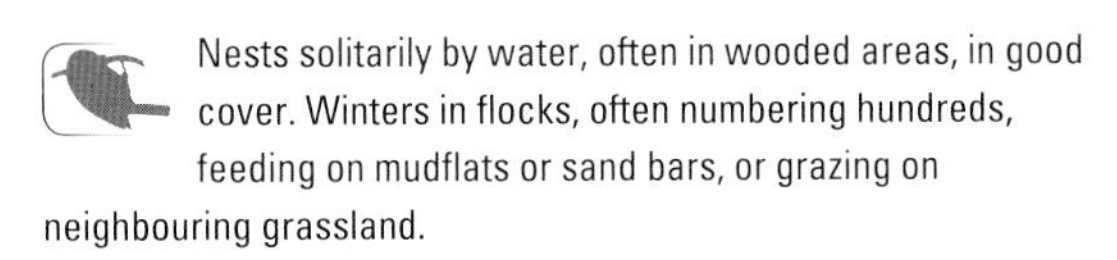

Nests solitarily by water, often in wooded areas, in good cover. Winters in flocks, often numbering hundreds, feeding on mudflats or sand bars, or grazing on neighbouring grassland.

TOP: *FEMALE*
ABOVE: *MALE*

FACT FILE

LENGTH 42–50cm WINGSPAN 71–85cm WEIGHT 500–900g FOOD Grass, algae, eelgrass EGGS 7–8, creamy buff HABITAT Winters on lakes, estuaries, deltas. Breeds by N lakes VOICE Male, diagnostic *whee-oo* whistles; female, purrs

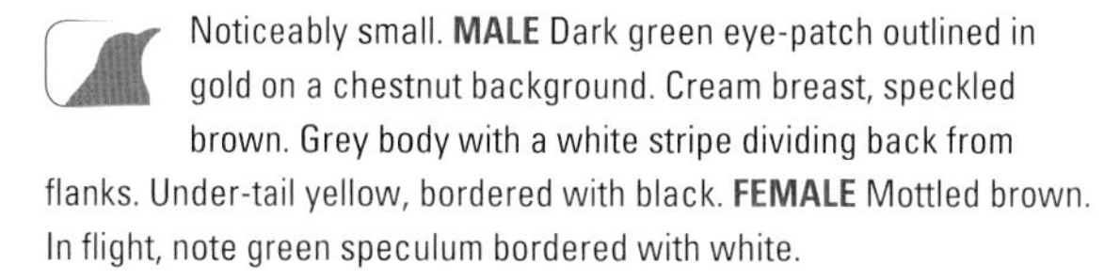

TEAL

Anas crecca

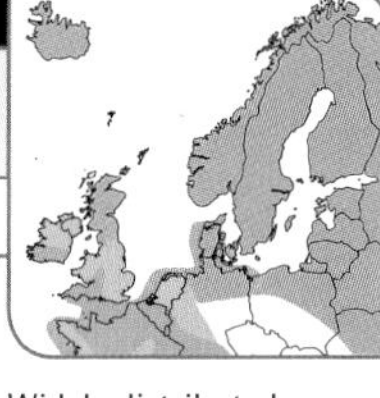

Noticeably small. **MALE** Dark green eye-patch outlined in gold on a chestnut background. Cream breast, speckled brown. Grey body with a white stripe dividing back from flanks. Under-tail yellow, bordered with black. **FEMALE** Mottled brown. In flight, note green speculum bordered with white.

Nests close to water. Winters in small flocks. Rises quickly from water and flies fast.

Widely distributed, breeding in Iceland, British Isles and France, eastwards across Asia. W populations winter in S Europe, N and E populations mainly in the Netherlands and British Isles.

LEFT: *FEMALE*
BELOW: *MALE*

FACT FILE

LENGTH 34–38cm WINGSPAN 53–59cm WEIGHT 240–360g FOOD Omnivorous; seeds especially in winter EGGS 8–10, pale stone to greenish buff HABITAT Breeds in marshes, wet upland moorland. Winters on lakes, coastal lagoons VOICE Male, melodious *prip-prip* calls; female, quacks

JAN	**FEB**	**MAR**	APR	MAY	JUN	JUL	AUG	SEP	**OCT**	**NOV**	**DEC**

GOLDENEYE

Bucephala clangula

Summer visitor to Scandinavia, Russia and across N Asia. Most winter in Baltic westwards to British Isles.

MALE Head glossy green with a white spot. White breast, belly, flanks, speculum and wing coverts; black back and tail end. **FEMALE** Mostly brown head with a pale yellow eye; rest of plumage mottled grey. In flight, shows less white on wing than male.

Nests in a tree-hole or nestbox up to 5m off ground. Ducklings tumble out, are led to water, then feed themselves. Dives for food.

RIGHT: *FEMALE*
BELOW: *MALE*

FACT FILE

LENGTH 40–48cm WINGSPAN 62–67cm WEIGHT 700–1,100g FOOD Aquatic animals EGGS 8–11, bluish green HABITAT Breeds in tall forest close to water. Winters on lakes, reservoirs, sheltered coastal waters VOICE Usually silent. Courtship calls: loud *zeeee-ZEEEE* and quieter *rrrrrrrrr*

JAN FEB MAR APR MAY JUN JUL AUG SEP OCT NOV DEC

POCHARD

Aythya ferina

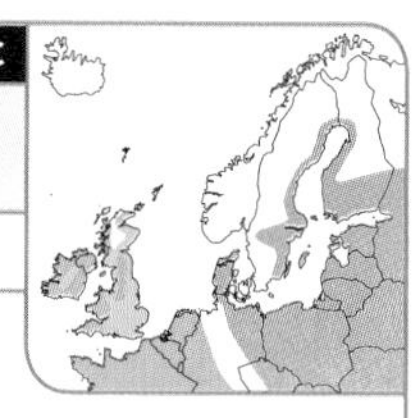

MALE Distinctive chestnut-red head, light grey back, and black breast and tail end. **FEMALE** Head and breast brown, back tinged grey; rest of underparts white with grey markings. Both sexes have a blackish bill with a broad, paler central band; in flight, they show a long greyish-white wing bar.

Nests in or very close to water on a mound of vegetation built by female. Dives for food.

Across British Isles to Baltic and into Asia. British population is augmented in winter by migrants from the E.

LEFT: *FEMALE*
BELOW: *MALE*

FACT FILE

LENGTH 42–49cm WINGSPAN 72–82cm WEIGHT 500–1,300g FOOD Mainly aquatic plants EGGS 8–10, green-grey HABITAT Breeds on lakes, large ponds and slow-flowing rivers with cover. Winters on open waters of lakes, reservoirs and estuaries, but rarely on open sea VOICE Male makes a courtship whisper and louder *kil-kil-kil*

JAN | FEB | MAR | APR | MAY | JUN | JUL | AUG | SEP | OCT | NOV | DEC

TUFTED DUCK

Aythya fuligula

Across the region, including Iceland. S birds are largely resident, but Icelandic and E birds migrate to milder S and W Europe.

MALE All black save for pure white flanks. **FEMALE** Mostly dark brown, with some white on sides and a little white at base of bill. Both have a long white bar on dark wings.

Nests on ground near water, even in gull and tern colonies. Tolerant of human presence on reservoirs and in city parks.

RIGHT: *FEMALE*
BELOW: *MALE*

FACT FILE

LENGTH 40–47cm WINGSPAN 65–72cm WEIGHT 500–740g FOOD Seeds, aquatic vegetation EGGS 8–10, pale buff HABITAT Breeds on freshwater lakes and rivers with fringes of good cover. Winters on open, sheltered waters VOICE Male, musical whistle and quiet *wek*, mostly during courtship

JAN	FEB	MAR	APR	MAY	JUN	JUL	AUG	SEP	OCT	NOV	DEC

EIDER

Somateria mollissima

Wholly coastal; roughly N of 55°N, across region from N England to Arctic.

ADULT MALE White with black crown, flanks, belly and tail. Sides of head are lime-green. **FEMALE** Brown with darker bars and mottling. **IMMATURE MALE** Patchy pied plumage.

Gregarious, often seen offshore for much of year. Young form crèches with female 'guardians' or 'aunties'. Protected for centuries in England, Iceland, Norway and elsewhere to maintain collection of nest down for sleeping bags and eiderdowns.

LEFT: *FEMALE*
BELOW: *MALE*

FACT FILE

LENGTH 60–70cm WINGSPAN 95–105cm WEIGHT 1.2–2.8kg FOOD Blue Mussels, other shellfish EGGS 4–6, greenish grey HABITAT Confined to marine habitats along rocky coasts VOICE Male, loud, cooing courtship call: *a-HOOO, A-HOOO*

JAN FEB MAR APR MAY JUN JUL AUG SEP OCT NOV DEC

GOOSANDER

Mergus merganser

Scattered breeding in UK S to Devon, Iceland, Baltic, Scandinavia and Russia. British and Icelandic birds are largely resident. Others migrate W to winter around Baltic, North Sea and Britain.

MALE Black back; white body, tinged pink; black head, glossed green. Bill long, fine, hook-tipped. White secondaries and coverts show in flight. **FEMALE** Chestnut head, sharply divided from white underparts; rest of body grey. In flight, shows white secondaries, grey coverts and blackish primaries.

Nests in a tree-hole, rock crevice or nestbox. In winter, sometimes seen in large flocks, even fishing communally.

RIGHT: *FEMALE*
BELOW: *MALE*

FACT FILE

LENGTH 58–68cm WINGSPAN 78–94cm WEIGHT 1.0–1.8kg FOOD Fish EGGS 8–11, creamy white HABITAT Breeds in hilly, forested country beside sheltered waters. Winters on large, open waters VOICE Quiet courtship calls. Alarm: harsh *garrrr*

JAN | FEB | MAR | APR | MAY | JUN | JUL | AUG | SEP | OCT | NOV | DEC

RED-BREASTED MERGANSER

Mergus serrator

MALE Black head, glossed green. White neck-ring; chestnut breast spotted with black, side of breast black with white spots, flanks and tail grey, back black, innerwing white, primaries dull black. **FEMALE** Reddish-brown head, fading into grey body. Crest less noticeable than male's.

Nests on ground in good cover.

Widespread from N British Isles and Iceland, eastwards across Scandinavia. In winter, E birds migrate W; others disperse from inland breeding sites.

LEFT: *FEMALE*
BELOW: *MALE*

FACT FILE

LENGTH 52–58cm WINGSPAN 67–82cm WEIGHT 900–1,100g FOOD Fish EGGS 8–10, stone to buff-olive HABITAT On the sea in winter. Breeds beside a lake or slow-flowing river near woodland VOICE Male, rough, drawn-out courtship purring. Alarm: gruff *graa*

Widespread to about 60°N. Resident in the W, but E birds migrate to within rest of range.

JAN FEB MAR APR MAY JUN JUL AUG SEP OCT NOV DEC

MOORHEN

Gallinula chloropus

ADULT Looks dull black at a distance, with a white line along its side; white under-tail sides very noticeable when it flicks its tail. Conspicuous red bill and basal shield, with a yellow tip. Yellowish-green legs and feet. **JUVENILE** Dull, dark brown, paler below.

Territorial. Feeds on the water, and in the open on grassland. Has 2–3 broods a year; earlier young help to feed new chicks.

RIGHT: *ADULTS FIGHTING*
BELOW: *ADULT*

FACT FILE

LENGTH 32–35cm WINGSPAN 50–55cm WEIGHT 182–365g FOOD Mostly vegetable matter EGGS 5–9; buff, variably marked with brown, black and purplish HABITAT Still or slow-flowing fresh waters, even on farms and city parks VOICE Call: *curruc*

JAN FEB MAR APR MAY JUN JUL AUG SEP OCT NOV DEC

COOT

Fulica atra

ADULT Slatey black save for white bill and frontal shield. In flight, shows a pale border to edge of wing. Greenish-grey legs and lobed toes. **JUVENILE** Back and wings duller than adult's, and face and underparts whitish.

For breeding, prefers larger waters than Moorhen (opposite). Very aggressive in territorial disputes, stabbing with bill, and kicking and beating with wings, all on the water.

Widespread; range very similar to Moorhen's. Winter visitors to Britain form flocks of >1,000 on several lakes and reservoirs.

FACT FILE

LENGTH 36–38cm WINGSPAN 70–80cm WEIGHT 610–1,200g FOOD Omnivorous, but mainly plant material EGGS 6–10; buff, evenly speckled dark brown HABITAT Lakes and reservoirs with marginal cover. Winters on large waters (even floods) with no cover VOICE Short, sharp, repeated calls, e.g. *kowk kowk*

JAN	FEB	MAR	**APR**	**MAY**	**JUN**	**JUL**	**AUG**	SEP	OCT	NOV	DEC

CRANE

Grus grus

Reintroduced to Britain in 2014, breeds, winter flock of 40–50. Summer visitor to Germany and Scandinavia eastwards. W populations migrate SW to winter mostly in S Spain; others migrate SE as far as Ethiopia.

20% larger than Grey Heron (p. 20). Long-necked and long-legged. Body and wings lead-grey, with long, drooping 'tail' (actually elongated secondary feathers), and contrasting head and neck. In flight, black wing-tips show; neck and legs are stretched out.

Shy and wary. Large flocks on migration and at spectacular dancing displays. Calls signal its presence.

FACT FILE

LENGTH 110–120cm WINGSPAN 220–245cm WEIGHT 3–7kg FOOD Mostly plant material EGGS 2; variable in colour, blotched brown HABITAT Breeds in swampy woodland, *Sphagnum* bogs, reedbeds. Winters in open wetlands, grassland, cultivation VOICE Loud, trumpeting calls

OYSTERCATCHER

Haematopus ostralegus

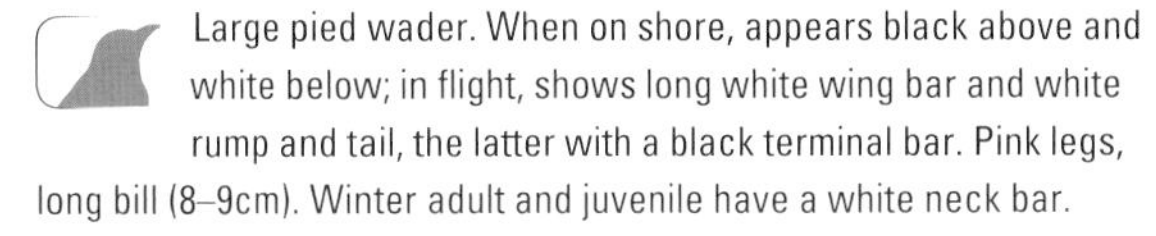

Large pied wader. When on shore, appears black above and white below; in flight, shows long white wing bar and white rump and tail, the latter with a black terminal bar. Pink legs, long bill (8–9cm). Winter adult and juvenile have a white neck bar.

Some nest inland on shingle by gently flowing rivers. Gregarious yet wary. Feeds by probing, and opens shells by hammering or stabbing with chisel-like bill. Noisy.

Iceland, British Isles and France, eastwards to Baltic and Scandinavia. N and E birds in particular migrate to milder W.

LEFT: *WINTER*
BELOW: *SUMMER*

FACT FILE

LENGTH 40–45cm WINGSPAN 80–86cm WEIGHT 400–700g FOOD Mainly bivalve shellfish EGGS 3; buffish, streaked and blotched blackish brown HABITAT Chiefly rocky, sandy or estuarine seashores VOICE Loud, shrill *kwik kwik kwik*, often much repeated

JAN | FEB | MAR | APR | MAY | JUN | JUL | AUG | SEP | OCT | NOV | DEC

AVOCET

Recurvirostra avosetta

SE England and N France, eastwards to S Baltic. Winters locally in Britain and the Netherlands, but more often from Iberia to coasts of Africa. Its presence depends on careful habitat management.

Has a unique combination of pied plumage, long blue-grey legs and slender, upturned bill. Mostly white with a black-capped head, and black wing-tips, coverts and sides of mantle.

Has a distinctive feeding method, sweeping its bill methodically from side to side with the open, curved tip just under shallow water, catching prey by touch.

FACT FILE

LENGTH 42–45cm WINGSPAN 77–80cm WEIGHT 230–395g FOOD Invertebrates EGGS 3–4; pale buff, variably marked black and grey HABITAT Extensive lowland coastal mudflats or brackish lagoons for feeding; nearby dry sandy or muddy flats or low islands for nesting VOICE Melodious, variable *klute-klute-klooit*

JAN | FEB | MAR | APR | MAY | JUN | JUL | AUG | SEP | OCT | NOV | DEC

RINGED PLOVER

Charadrius hiaticula

ADULT Back brown. Underparts white with a broad black breast-ring. Head patterned black–brown–white. Wing coverts brown, flight feathers black with white bar right across. Tail brown in centre with white edges. **JUVENILE** Lacks black on head; breast-band broken.

Energetic; runs a few metres, stops and bends to feed, then runs again, and so on.

Mostly coastal in Iceland and British Isles to Scandinavia. British birds are largely resident; E European and Scandinavian birds winter widely from British Isles to W Africa.

FACT FILE

LENGTH 18–20cm WINGSPAN 48–57cm WEIGHT 61–65g FOOD Invertebrates EGGS 3–4; variable, with few or many black markings HABITAT Breeds on sandy or pebbly shores, and dry sandy or gravelly land by lakes or rivers inland. Winters mainly on sandy or muddy shores and estuaries VOICE Liquid *tooLEE*

JAN | FEB | MAR | APR | MAY | JUN | JUL | AUG | SEP | OCT | NOV | DEC

GOLDEN PLOVER

Pluvialis apricaria

Breeds in Iceland, N British Isles and across Scandinavia. Migrates to W Europe, and S to Iberia and N Africa.

Upperparts distinctively spangled black and gold; tail barred. **SUMMER ADULT** Underparts black, divided from upperparts by white band down side. N birds are more intensely black. **WINTER ADULT and IMMATURE** Underparts mottled grey-brown, fading to white on belly; off-white supercilium; white underwing.

Breeds chiefly in heather or moor grass. In winter, forms flocks to feed and roost off the moors.

BELOW: *ADULT SUMMER SOUTHERN RACE;* RIGHT: *JUVENILE*

N birds have black faces outlined in white

FACT FILE

LENGTH 26–29cm WINGSPAN 67–76cm WEIGHT 175–275g FOOD Invertebrates EGGS 4; variable ground colour, heavily marked HABITAT Breeds on upland moors, peatlands. Winters on grasslands, stubble and arable fields and, in hard weather, mudflats VOICE Call: musical *tlui* whistle

LAPWING

Vanellus vanellus

ADULT Black and white patterned head and diagnostic crest; black breast, white belly; pale orange under tail. Black back. Wings broad and rounded, black with white tips. Tail black and white. **JUVENILE** Shorter crest and pale feather edges.

Male has a spectacular aerial display in spring. Winter flocks have a 'leisurely' flight, with slow wingbeats.

Widespread across region, but not Iceland. Mainly a summer visitor, wintering in British Isles and W maritime Europe. Very susceptible to cold weather, moving on to avoid it.

FACT FILE

LENGTH 28–31cm WINGSPAN 82–87cm WEIGHT 185–250g FOOD Invertebrates EGGS 4; olive or clay-coloured, blotched black HABITAT Mainly open arable ground; also favours rushy fields and moorland for breeding. Hard weather drives flocks to muddy estuaries VOICE Call: variations on *PEE- wit*

JAN FEB MAR APR MAY JUN JUL AUG SEP OCT NOV DEC

SANDERLING

Calidris alba

Breeds in Spitzbergen and Siberia, and Arctic Canadian islands and Greenland. All populations are passage migrants and winter visitors to Denmark westwards, and onwards to Africa. Early arrivals are in summer plumage.

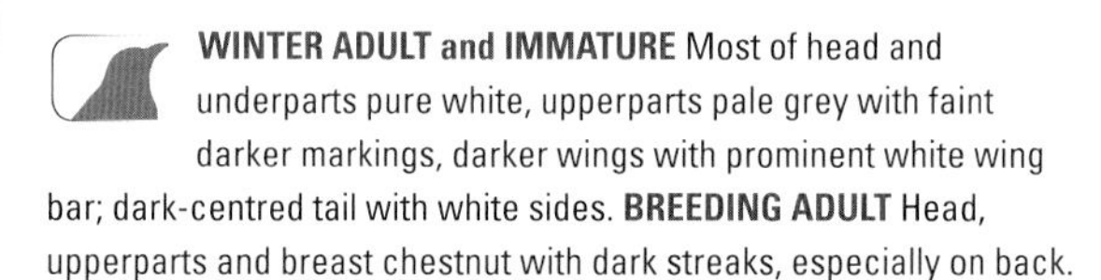

WINTER ADULT and IMMATURE Most of head and underparts pure white, upperparts pale grey with faint darker markings, darker wings with prominent white wing bar; dark-centred tail with white sides. **BREEDING ADULT** Head, upperparts and breast chestnut with dark streaks, especially on back.

Rushes along in small flocks, legs a blur. Feeds at water's edge, snatching sandhoppers.

FACT FILE

LENGTH 20–21cm WINGSPAN 40–45cm WEIGHT 51–60g FOOD Small invertebrates EGGS 4; dull olive, sparsely spotted brown HABITAT Breeds on high-Arctic tundra. Migrating and wintering birds frequent sandy shores VOICE Flight call outside breeding season: soft *twick twick*

JAN | FEB | MAR | APR | MAY | JUN | JUL | AUG | SEP | OCT | NOV | DEC

DUNLIN

Calidris alpina

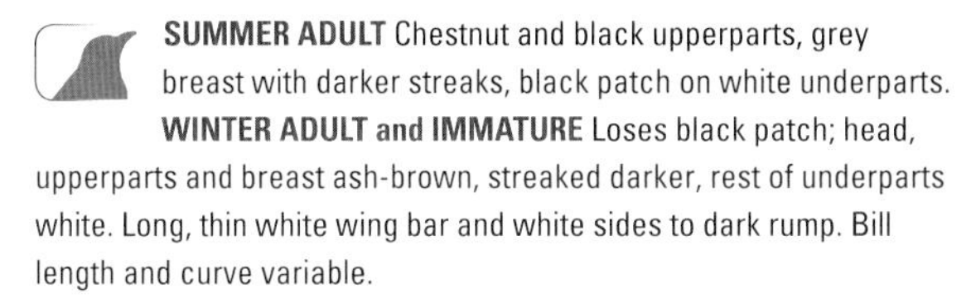

SUMMER ADULT Chestnut and black upperparts, grey breast with darker streaks, black patch on white underparts. **WINTER ADULT and IMMATURE** Loses black patch; head, upperparts and breast ash-brown, streaked darker, rest of underparts white. Long, thin white wing bar and white sides to dark rump. Bill length and curve variable.

Dartmoor, England, is the most S site in the world. Often winters in large flocks.

Breeds in Iceland, British Isles and Scandinavia, with some S of Baltic. Winters in British Isles and S to NW Africa.

LEFT: *WINTER*
BELOW: *SUMMER*

FACT FILE

LENGTH 16–20cm WINGSPAN 38–43cm WEIGHT 41–56g FOOD Chiefly invertebrates EGGS 4; buff or olive, marked with brown blotches HABITAT Nests on high moorlands with boggy areas, and tussocky tundra. Winters on muddy and sandy shores and estuaries VOICE Shrill, slurred *treeep*

JAN FEB MAR APR MAY JUN JUL AUG SEP OCT NOV DEC

TURNSTONE

Arenaria interpres

Breeds in a coastal band around Scandinavia and Baltic, and all around the Arctic. Canadian and Greenland birds winter in W Europe; our birds migrate to NW and W Africa.

BREEDING ADULT Black and white patterned head, black breast, and black and chestnut upperparts. Short orange-yellow legs. Unique flight pattern: white shoulders, wing bars, lower back and base of tail; the rest looks black. **WINTER ADULT and IMMATURE** Much plainer (*see* photo).

Well named – actively turns over stones and seaweed. On sandy shores, digs for sandhoppers above the high-tide mark. Rather quarrelsome in feeding flocks.

TOP: *WINTER;* ABOVE: *SUMMER*

FACT FILE

LENGTH 22–24cm WINGSPAN 50–57cm WEIGHT 105–160g FOOD Insects when breeding; invertebrates at other times EGGS 4; olive, heavily marked with dark brown HABITAT Rocky, seaweedy shores VOICE Flight call: distinctive twittering *kitititit*

JAN FEB MAR APR MAY JUN JUL AUG SEP OCT NOV DEC

COMMON SANDPIPER

Actitis hypoleucos

N and W British Isles, Baltic and Scandinavia; scattered elsewhere. A few winter in mild W Europe, most in tropical Africa.

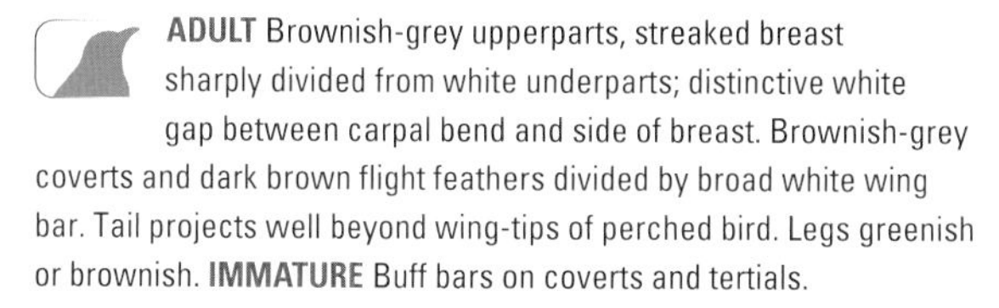

ADULT Brownish-grey upperparts, streaked breast sharply divided from white underparts; distinctive white gap between carpal bend and side of breast. Brownish-grey coverts and dark brown flight feathers divided by broad white wing bar. Tail projects well beyond wing-tips of perched bird. Legs greenish or brownish. **IMMATURE** Buff bars on coverts and tertials.

Walks with a wagtail-like bobbing. Unique wing action: flicked wingbeats, then a glide, on wings held below shoulder level.

LEFT: *ADULT*
BELOW: *JUVENILE*

FACT FILE

LENGTH 19–21cm WINGSPAN 38–41cm WEIGHT 40–59g FOOD Invertebrates, especially insects EGGS 4, pale buff with red-brown markings HABITAT Breeds by hill streams or lakes. On migration, stops by lowland waters and saltmarshes VOICE Shrill *seep-seep-seep*

JAN FEB MAR APR MAY JUN JUL AUG SEP OCT NOV DEC

REDSHANK

Tringa totanus

Iceland, British Isles, Low Countries, around Baltic, and W Scandinavia. Most W birds winter in British Isles; others head to Mediterranean and W Africa.

BREEDING ADULT Olive-brown upperparts, marked with black spots; white underparts, with breast and flanks spotted and streaked dark. In flight shows conspicuous white rump, tail coverts and wing bars. **WINTER ADULT** Plainer.

Male guards territory from a viewpoint on a post. Gregarious on passage nd in winter.

RIGHT: *WINTER*
BELOW: *SUMMER*

FACT FILE

LENGTH 27–29cm WINGSPAN 59–66cm WEIGHT 105–165g FOOD Worms, insects in summer; crustaceans, molluscs, worms in winter EGGS 4, cream with red-brown spots HABITAT Breeds on marshes and wet grasslands in lowlands and valleys. Winters on muddy estuaries and creeks VOICE Ringing *TEU-hu-hu*, the last 2 notes lower in pitch

JAN	FEB	MAR	APR	MAY	JUN	JUL	AUG	SEP	OCT	NOV	DEC

GREENSHANK

Tringa nebularia

Summer visitor to N Scotland, Scandinavia and N Russia. Scottish birds winter in Ireland and W Britain, the rest in tropical Africa.

NON-BREEDING ADULT and IMMATURE Grey above, white below, with greyish streaks on head, neck and breast. In flight, wings look dull black, in contrast to white lower back, rump and barred tail. Bill dark grey, legs olive-green. **SUMMER ADULT** Similar, but black spots on back and flanks.

Nest is on ground close to a marker – a stone, log or tree stump.

BELOW: *ADULT WINTER;* BOTTOM: *JUVENILE*

FACT FILE

LENGTH 30–33cm WINGSPAN 68–70cm WEIGHT 130–260g FOOD Small fish and invertebrates EGGS 4; off-white, marked with red and brown HABITAT N Scotland's treeless moorlands, lochs and bogs. Elsewhere, mountain valleys with bogs, lakes, pools and scattered trees VOICE Ringing *tew, tew, tew*

JAN | FEB | MAR | APR | MAY | JUN | JUL | AUG | SEP | OCT | NOV | DEC

BLACK-TAILED GODWIT

Limosa limosa

Breeds commonly in Iceland, a few in Britain and many in Germany eastwards below 60°N. Icelandic birds winter in British Isles; others winter in Africa, mainly N of Equator.

Tall wader. **SUMMER ADULT** Unique flight pattern: brown coverts, broad white bar across black secondaries and primaries, white lower rump and base of tail, and black terminal tail band. Legs project well beyond tail. **WINTER ADULT and JUVENILE** Coloured plumage becomes plain, pale grey.

WINTER

Breeds in dispersed colonies. Sometimes wades deeply to feed. Very gregarious in non-breeding flocks.

FACT FILE

LENGTH 40–44cm WINGSPAN 70–82cm WEIGHT 209–425g FOOD Chiefly invertebrates EGGS 3–4; olive or brown, blotched darker HABITAT Breeds on marshy moorland, rough pasture, water meadows; winters on mudflats, estuaries, freshwater marshes VOICE Flight call: *wicka-wicka-wicka*

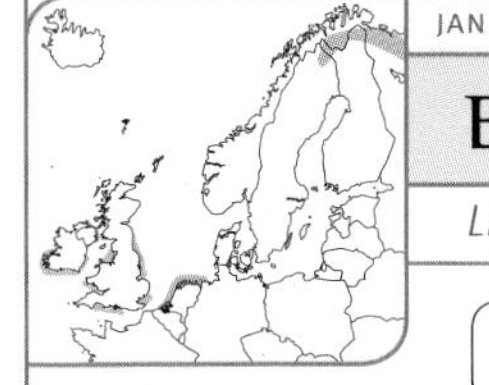

JAN | FEB | MAR | **APR** | **MAY** | JUN | JUL | **AUG** | **SEP** | **OCT** | NOV | DEC

BAR-TAILED GODWIT

Limosa lapponica

Summer visitor to 65–70°N in Fennoscandia. Passage migrant and winter visitor to British Isles, the Netherlands and Germany.

Shorter-legged than Black-tailed, with upcurved bill, no obvious wing bar, and black- and white-barred tail. **SUMMER ADULT** Head, neck and underparts chestnut-red; upperparts grey, mottled black and rufous. **WINTER ADULT and JUVENILE** Grey head; upperparts grey, streaked white; underparts mostly white.

WINTER

Territorial breeder; later, very gregarious in large flocks.

FACT FILE

LENGTH 37–39cm WINGSPAN 70–80cm WEIGHT 228–419g FOOD Chiefly invertebrates EGGS 3–4; green to olive, lightly speckled brown and grey HABITAT Breeds on peat, heath tundra; otherwise coastal, on sand or mud VOICE Call: *kirruc, kirruc*

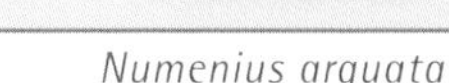

CURLEW

Numenius arquata

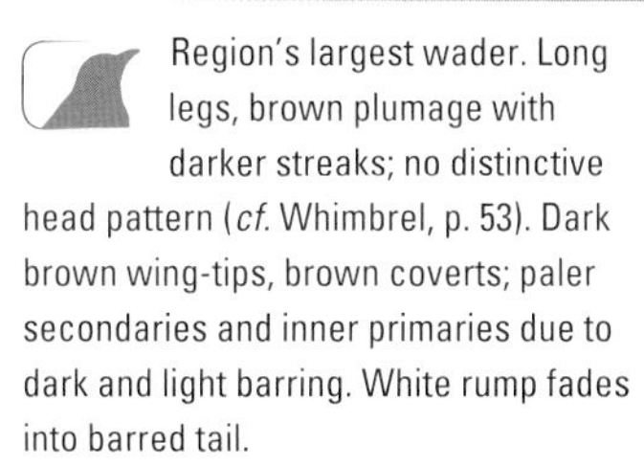

Region's largest wader. Long legs, brown plumage with darker streaks; no distinctive head pattern (*cf.* Whimbrel, p. 53). Dark brown wing-tips, brown coverts; paler secondaries and inner primaries due to dark and light barring. White rump fades into barred tail.

Probes deeply for invertebrates. The most wary of waders, but gregarious, often in large flocks.

Breeds widely across the region except Iceland and the central spine of Scandinavia. Winters in the W, southwards to Africa. Serious decline in Europe.

FACT FILE

LENGTH 50–60cm WINGSPAN 80–100cm WEIGHT 600–1,100g FOOD Omnivorous EGGS 4, green to olive with small darker spots HABITAT Breeds on moist moorland, rough grasslands, damp pastures; later, favours mostly mudflats, saltings, and neighbouring grassland at high tide VOICE Characteristic call, heard all year: *CUR-lee, CUR-lee*, often running into a rich, bubbling sound

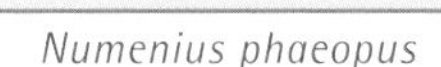

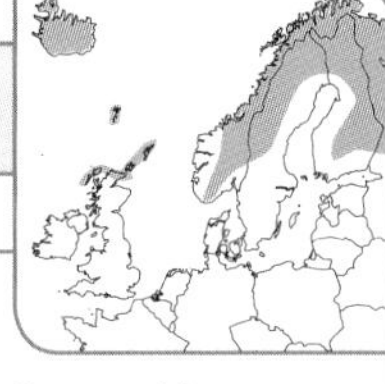

WHIMBREL

Numenius phaeopus

Dark brown crown with buff centre stripe, buff supercilium, dark eye stripe, and pale cheeks and throat. Most of plumage mottled dark brown, belly and under-tail whitish. No wing bar, but lower back and rump white. Bill is like that of a small Curlew (p. 53).

Gregarious on migration, usually in small flocks. On migration calls a rapid, evenly paced *titititititi.*

Summer visitor to Iceland, far N Scotland and islands, Scandinavia and Russia; passage migrant to W Europe; winters in Africa S to Cape of Good Hope.

FACT FILE

LENGTH 40–42cm WINGSPAN 76–89cm WEIGHT 430–575g FOOD Omnivorous EGGS 4; green to olive with small, darker markings HABITAT Breeds on moorlands, heaths, by tundra pools; later, is coastal, on muddy and sandy beaches, and rocky shores VOICE See above

JAN FEB MAR APR MAY JUN JUL AUG SEP OCT NOV DEC

SNIPE

Gallinago gallinago

Breeds widely across region. British Isles population disperses locally. Other populations migrate to British Isles and NW Europe.

Medium-sized wader. Head dark brown through eye, then buff and dark below, brighter buff and dark above, and dark crown bisected with buff line. Dark, richly patterned upperparts, beautiful gold edges to black and chestnut scapular feathers. Legs greenish.

In spring, male dives over territory making a 'drumming' or 'bleating' sound as outer tail feathers vibrate. Probes for food.

FACT FILE

LENGTH 25–27cm WINGSPAN 44–47cm WEIGHT 90–130g FOOD Chiefly invertebrates EGGS 4; olive with small, dark blotches HABITAT Open wet ground, marshes, bogs and peat moors with soft soil VOICE Persistent spring call: *CHIPper, CHIPper, CHIPper*. Towers when flushed, making a harsh call: *scaaap*

JAN FEB **MAR APR MAY JUN JUL AUG SEP OCT** NOV DEC

RUFF

Philomachus pugnax

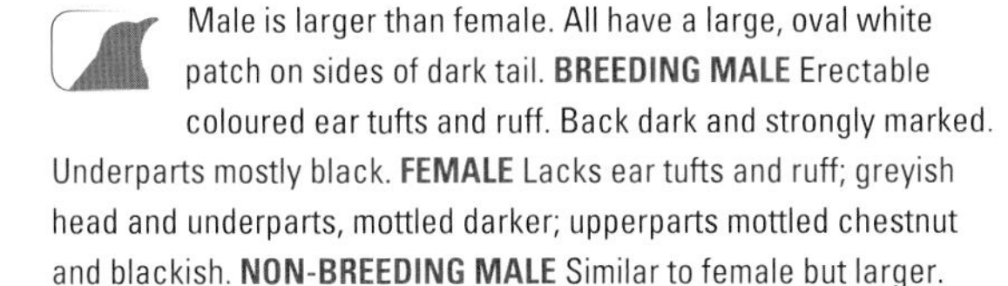

Male is larger than female. All have a large, oval white patch on sides of dark tail. **BREEDING MALE** Erectable coloured ear tufts and ruff. Back dark and strongly marked. Underparts mostly black. **FEMALE** Lacks ear tufts and ruff; greyish head and underparts, mottled darker; upperparts mottled chestnut and blackish. **NON-BREEDING MALE** Similar to female but larger.

Scarce summer visitor across region. Some winter in W Europe but most in sub-Saharan Africa.

5–20 males display in a traditional area. Dominant males hold territory there and successfully mate with several females.

RIGHT: *FEMALE*
BELOW: *MALE*

FACT FILE

LENGTH 20–30cm WINGSPAN 48–58cm WEIGHT 80–230g FOOD Invertebrates EGGS 4; greenish with small, dark brown marks HABITAT Breeds on wetlands with pools and low vegetation. Winters on muddy margins of waters VOICE Mostly silent

JAN FEB MAR **APR MAY JUN JUL AUG SEP OCT** NOV DEC

ARCTIC SKUA

Stercorarius parasiticus

Summer visitor to Iceland, far N Scotland and islands, coastal Scandinavia and Russia. Pelagic outside breeding season into N and S Atlantic.

All have white wing-patches. Adults have 2 long, central tail feathers and 2 colour phases. **LIGHT-PHASE ADULT** Black cap; white cheeks, yellowish at back; light grey breast-band, rest of underparts white; brown-grey upperparts. **DARK-PHASE ADULT** All sooty grey-brown, save for darker cap. Intermediates also occur. **JUVENILE** Dark bars below and rufous-barred upperparts.

Colonial. Aggressive against intruders. Swift and agile in pursuit of gulls, terns and auks for their fish.

RIGHT: *DARK PHASE*
BELOW: *LIGHT PHASE*

FACT FILE

LENGTH 41–46cm WINGSPAN 110–125cm WEIGHT 360–590g FOOD Small mammals, birds, eggs. At sea, mainly fish EGGS 2, olive with brownish markings HABITAT Nests on tundra, moorland, islands. Non-breeding birds pelagic VOICE *Kee-ah* and *tuk*

JAN | FEB | MAR | APR | MAY | JUN | JUL | AUG | SEP | OCT | NOV | DEC

BLACK-HEADED GULL

Larus ridibundus

SUMMER ADULT Chocolate-brown (not black) hood, pale blue-grey upperparts, white body and tail. Wings tipped black; underside of primaries dark grey. Bill and legs red. **WINTER ADULT** No hood, just dark spot behind eye. **IMMATURE** Yellowish bill and legs, black trailing edge to wing behind white panel, black-tipped tail.

Colonial. Often seen following the plough.

Breeds in Iceland and British Isles, eastwards to Russia. E and N populations migrate S and W. Others disperse as far as Mediterranean.

diagnostic broad white band at front of primaries

TOP: *WINTER;* ABOVE: *SUMMER*

FACT FILE

LENGTH 34–37cm WINGSPAN 100–110cm WEIGHT 160–310g FOOD Mainly insects and worms; also scavenges EGGS 2–3, greenish or brown with darker spots HABITAT Breeds on sand-dunes, and beside lakes and marshes. Winters on estuaries, playing fields, lakes in parks, reservoirs VOICE Distinctive laughing call

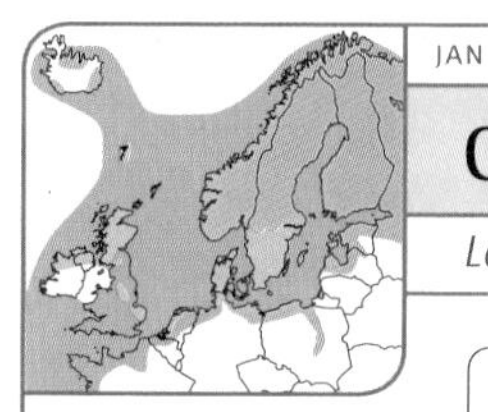

JAN | FEB | MAR | **APR** | **MAY** | **JUN** | **JUL** | **AUG** | **SEP** | **OCT** | NOV | DEC

COMMON GULL

Larus canus

A few in Iceland, British Isles (mostly Scotland and N Ireland) and Low Countries, eastwards to Russia and beyond. Winters on Baltic, and W Atlantic seaboard southwards to Brittany.

Smallish bill and rounded head. **SUMMER ADULT** White head and body, darkish grey wings with black primaries and large white tips. **WINTER ADULT** Speckled head. **IMMATURE** Almost black primaries, pale coverts, blackish bar across trailing edge of secondaries; white tail has black sub-terminal band. Gains adult plumage in 3rd year.

Colonial. Not the commonest gull, despite its English name.

RIGHT: *SUMMER*
BELOW: *WINTER*

FACT FILE

LENGTH 40–47cm WINGSPAN 110–130cm WEIGHT 290–530g FOOD Fish, marine and land invertebrates EGGS 3, olive or blue-green with brown markings HABITAT Nests by lochs and inshore inlets, on moorland and marshes. After breeding, on estuaries, grasslands, sports fields VOICE High-pitched *keee-ya*

HERRING GULL

Larus argentatus

ADULT Upperparts pale grey, with white leading and trailing edges to wings; black wing-tips with separated white spots. Rest of plumage white. **FIRST-WINTER** Dusky, mottled brown with dark brown wing-tips and tail tip. Gains adult plumage in 5 years.

The typical 'seagull'. Very bold: will steal food from seaside tourists.

Widespread, common breeder on all coasts. Far N and E populations migrate S and W; others disperse to local estuaries and ports.

FACT FILE

LENGTH 55–67cm WINGSPAN 138–155cm WEIGHT 800–1,100g FOOD Omnivorous – almost anything! EGGS 2–3, olive with many dark brown spots and blotches HABITAT Nests on coastal cliffs, islands; recently also buildings. Non-breeders on estuaries, refuse dumps, fish quays, and following ships close to shore VOICE Loud *kyou-kyou-kyou*

JAN FEB **MAR APR MAY JUN JUL AUG SEP OCT NOV** DEC

LESSER BLACK-BACKED GULL

Larus fuscus

Breeds in Iceland, British Isles, Scandinavia and Baltic. Migrates S to Mediterranean and coastal NW Africa.

ADULT White head, body, tail and underwings. Mantle and upperwings dark grey, wing-tips black with white primary tips. **JUVENILE** Streaked grey-brown body and wings. At all ages and plumages, similar to Great Black-backed (p. 61). Gains adult plumage in 4th year.

Colonial ground-nester; some on rooftops of coastal cities.

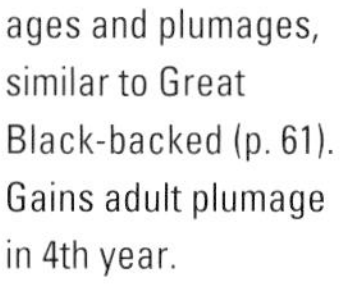

FACT FILE

LENGTH 48–56cm WINGSPAN 117–134cm WEIGHT 770–1,000g
FOOD Omnivorous, even rubbish EGGS 3, olive or brown with darker markings
HABITAT Nests on coastal and lake islets VOICE Calls: *kaw* and *ga-ga-ga*

JAN | FEB | MAR | APR | MAY | JUN | JUL | AUG | SEP | OCT | NOV | DEC

GREAT BLACK-BACKED GULL

Larus marinus

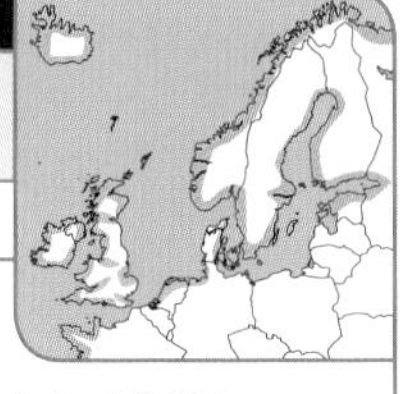

Iceland, British Isles, Brittany and Denmark eastwards. Most resident, simply dispersing to winter along region's coasts and off shore. Far NE populations migrate W and S.

ADULT Very large, with all-black back and wings, the latter with white tips and trailing edges. Rest of plumage white. **IMMATURE** Mottled and streaked brown but with paler head. Adult plumage gained after 5 years.

Serious breeding season predator of shearwaters and Puffin (p. 68), many caught in mid-air. In winter, scavenges at ports and rubbish dumps.

FACT FILE

LENGTH 64–78cm WINGSPAN 150–165cm WEIGHT 1–2.3kg FOOD Omnivorous EGGS 2–3; olive-brown, blotched darker brown HABITAT Nests on small islands and stacks; coastal in winter VOICE Deep, barking *aouk, aouk*

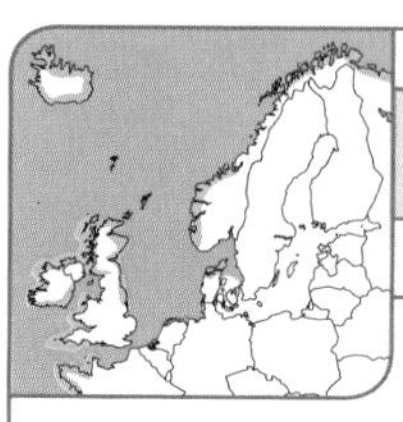

JAN FEB **MAR APR MAY JUN JUL AUG SEP OCT** NOV DEC

KITTIWAKE

Rissa tridactyla

Scattered cliff colonies on almost all coasts except Baltic. Winters S to about 35°N, even weathering huge storms at sea.

Rounded head, dark eye. **ADULT** Head and underparts white. Upperparts blue-grey with black-tipped wings (no spots). **JUVENILE** White head, body and tail; black spots above and behind eye, black rear neck band and terminal tail band; black wing-tips, bill, legs and diagonal wing bar; grey forewing and white behind bar.

In summer, may be seen miles offshore, feeding on or near the surface at a good food source. Easily recognised by its call.

FACT FILE

LENGTH 38–40cm WINGSPAN 95–120cm WEIGHT 305–525g FOOD Fish, invertebrates EGGS 1–3; pale buff, lightly spotted dark brown HABITAT Breeds on high, steep sea cliffs. Pelagic in non-breeding season VOICE *Kitti-wa-a-a-k*

JAN	FEB	**MAR**	**APR**	**MAY**	**JUN**	**JUL**	**AUG**	**SEP**	**OCT**	NOV	DEC

SANDWICH TERN

Sterna sandvicensis

Larger and whiter looking than other terns, with relatively longer wings. **SUMMER ADULT** Black cap and nape; upperparts pale ash-grey, with no dark wing-tip. Underparts white. Tail forked and short. **WINTER ADULT** White forehead (often gained before leaving breeding area). **IMMATURE** As winter adult but with darker wing-tips.

Nests colonially in sandy scrapes. Coasts are prone to erosion, so colonies move. Adult often seen on passage accompanied by squeaking juvenile.

Summer visitor from British Isles and Brittany to S Baltic. Winters off Africa, as far S as Cape of Good Hope.

FACT FILE

LENGTH 36–41cm WINGSPAN 95–105cm WEIGHT 225–285g FOOD Fish, especially sandeels EGGS 1–2, creamy with dark brown marks HABITAT Sandy, shingly coasts VOICE Loud, grating *kirrik*

JAN FEB MAR **APR MAY JUN JUL AUG SEP OCT** NOV DEC

COMMON TERN

Sterna hirundo

Not Iceland or forested highlands; otherwise scattered throughout region. Summer visitor, wintering especially on coasts of W Africa.

Long (7–12cm), forked tail. **SUMMER ADULT** Black cap; white underparts, rump and tail (greyish tint); pale grey upperparts with dark wedge to wing-tips. **WINTER ADULT** Forehead and crown white, bill black with reddish base, wings with dusky leading edge. **JUVENILE** Dark primaries, leading edge and bar across secondaries; black bill and legs.

Colonial breeder. Nest is a scrape on the ground; sometimes even far inland up rivers. Fishes by plunge-diving.

FACT FILE

LENGTH 31–35cm WINGSPAN 77–98cm WEIGHT 90–165g FOOD Chiefly marine fish EGGS 1–3; buff, spotted and streaked blackish brown HABITAT Shingle or sandy shores, coastal heaths, gravel pits VOICE High-pitched *keee-yah* and *kik-kik-kik*

JAN FEB MAR **APR MAY JUN JUL AUG SEP OCT** NOV DEC

ARCTIC TERN

Sterna paradisaea

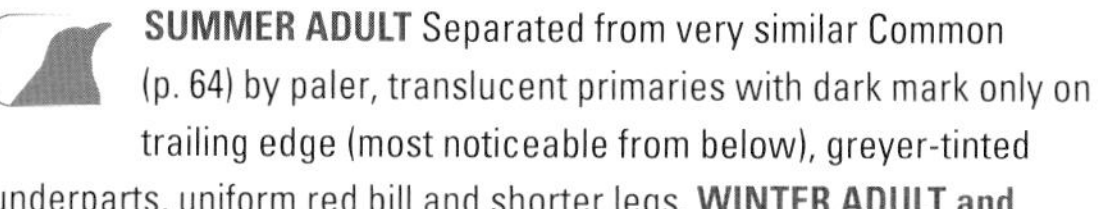

SUMMER ADULT Separated from very similar Common (p. 64) by paler, translucent primaries with dark mark only on trailing edge (most noticeable from below), greyer-tinted underparts, uniform red bill and shorter legs. **WINTER ADULT and IMMATURE** Very like winter Common but with all-black bill.

Has the longest migration of any bird. Nest sites similar to Common's. Aggressive against intruders, including humans.

Iceland, N British Isles and the Netherlands, eastwards across region and around Arctic. Summer visitor. No other tern breeds in high Arctic. Winters in S oceans, to the pack ice.

FACT FILE

LENGTH 33–35cm WINGSPAN 75–85cm WEIGHT 87–145g FOOD Marine fish, insects EGGS 1–3; pale buff, spotted dark brown HABITAT Mostly coastal, but also many kilometres up some rivers VOICE *Kee-arr*, higher in pitch than Common's

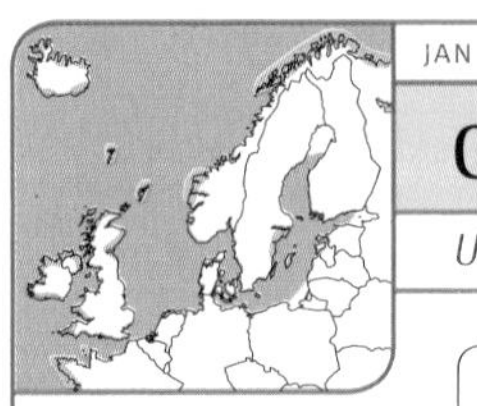

JAN | FEB | MAR | APR | **MAY** | **JUN** | **JUL** | **AUG** | SEP | OCT | NOV | DEC

GUILLEMOT

Uria aalge

Breeds in N and W British Isles, Iceland and coasts of Scandinavia. First-year birds disperse furthest, but no further than Portugal.

SUMMER ADULT Head, neck and upperparts chocolate-brown (looks black in poor light); underparts and trailing edge to secondaries white. Some birds (so-called 'bridled') show a white 'spectacle' around the eye. **WINTER ADULT and JUVENILE** Neck and face below and behind eye white.

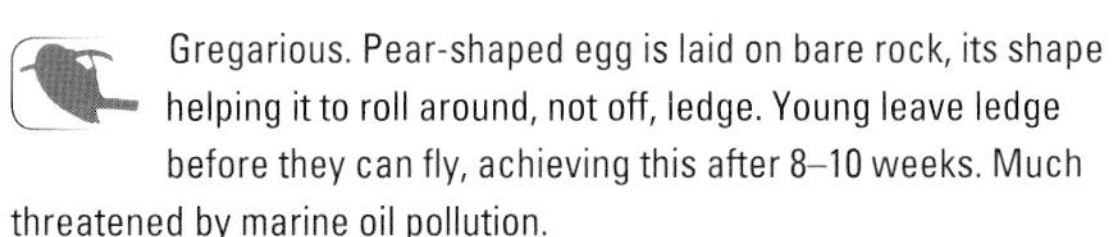

Gregarious. Pear-shaped egg is laid on bare rock, its shape helping it to roll around, not off, ledge. Young leave ledge before they can fly, achieving this after 8–10 weeks. Much threatened by marine oil pollution.

ABOVE: *BRIDLED FORM SUMMER*

FACT FILE

LENGTH 38–41cm

WINGSPAN 64–70cm

WEIGHT 650–1,300g

FOOD Mainly fish

EGGS 1; variable colour, scribbled and blotched

HABITAT Sea cliffs. Winters offshore

VOICE Growling and crowing

JAN FEB MAR APR **MAY JUN JUL AUG** SEP OCT NOV DEC

RAZORBILL

Alca torda

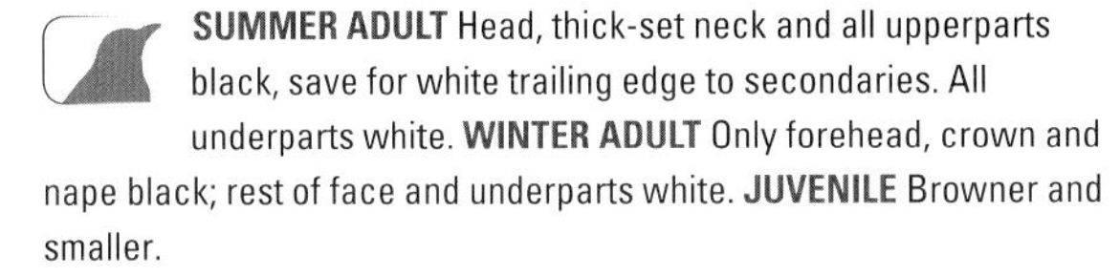

SUMMER ADULT Head, thick-set neck and all upperparts black, save for white trailing edge to secondaries. All underparts white. **WINTER ADULT** Only forehead, crown and nape black; rest of face and underparts white. **JUVENILE** Browner and smaller.

Often found with Guillemots (p. 66) on breeding cliffs, but lays egg on a wider ledge under an overhang, in a crevice or among boulders at base of cliff. Chick jumps to sea when only two-thirds grown, and is cared for by male.

Breeds in Brittany (a few), Faeroes, British Isles, Iceland and Scandinavia, including Baltic. N birds and juveniles migrate furthest, as far as Morocco.

FACT FILE

LENGTH 37–39cm
WINGSPAN 63–68cm
WEIGHT 600–800g
EGGS 1, brown or greenish with dark brown markings
HABITAT Sea cliffs. Winters offshore
VOICE Growling

JAN	FEB	MAR	**APR**	**MAY**	**JUN**	**JUL**	**AUG**	SEP	OCT	NOV	DEC

PUFFIN

Fratercula arctica

Breeds in Iceland, N and W British Isles, N coasts of Norway and Finland, and Spitzbergen. Wanders at sea in winter as far S as the Canaries and W Mediterranean.

All auks stand upright but look long and low when swimming. **SUMMER ADULT** Black above, white below; wings dark below. Huge bill (*see* photo). **WINTER ADULT** Bill reduced in size and colour, face greyish.

Lives in colonies. Nests in a burrow 1–2m long, often used again in other years. Several fish at a time brought to young in bill. 'Flies' underwater using wings.

FACT FILE

LENGTH 26–29cm
WINGSPAN 47–63cm
WEIGHT 310–560g
FOOD Fish; chicks are fed sandeels
EGGS 1, white
HABITAT Breeds on islands, sea cliffs
VOICE Creaking and growling

WILLOW GROUSE/RED GROUSE

Lagopus lagopus lagopus/Lagopus lagopus scoticus

BREEDING MALE WILLOW Upperparts and head rich red-brown, barred darker; wings and underbody white; tail black. **BREEDING FEMALE WILLOW** Barring more obvious and chin white. **WINTER WILLOW** Both sexes all white except black tail. **MALE RED** Never has white wings; entire plumage is rich red-brown, except for white underwing coverts and variable amount of white on underparts in winter. **FEMALE RED** Paler than male.

Red Grouse have long been husbanded on grouse moors. Call is often heard.

WILLOW GROUSE (above) resident across Scandinavia and into Asia. RED GROUSE (below) resident in W British Isles.

LEFT: *MALE WILLOW GROUSE*
BELOW: *MALE RED GROUSE*

FACT FILE

LENGTH BOTH 37–42cm
WINGSPAN 55–66cm
WEIGHT 600–795g
FOOD Almost entirely vegetarian
EGGS 6–9; pale yellow, mottled dark red-brown
HABITAT WILLOW breeds in tundra, moors, dwarf willow, birch. RED confined to heather moors
VOICE Loud *go-back go-back back back*

Scotland, Scandinavia, around Baltic. Decreasing in many areas from deforestation, disturbance and shooting.

JAN | FEB | MAR | APR | MAY | JUN | JUL | AUG | SEP | OCT | NOV | DEC

CAPERCAILLIE

Tetrao urogallus

MALE Blackish body and tail, brown wings; black is tipped white on shoulders and upper-tail coverts. **FEMALE** Brown, barred and mottled buff, black and white; rufous-orange breast patch, rounded tail.

Feeds in conifer crowns Oct–Apr; mostly on ground May–Sep. Ground-nesting, well-camouflaged female tends chicks alone. Male's extraordinary song, performed with raised tail, lasts *c.* 5 seconds and comprises several sounds running into each other: quiet tapping, getting faster, then pop, and ending with a grinding noise.

TOP: *FEMALE;* ABOVE: *MALE*

FACT FILE

LENGTH 60–87cm (male *c.* 40% larger) WINGSPAN 87–125cm WEIGHT 1.5–6.5kg FOOD Conifer shoots and buds, berries, other vegetable matter EGGS 2–10; pale buff, lightly speckled red-brown HABITAT Coniferous forest with glades VOICE *See* Habits

JAN | FEB | MAR | APR | MAY | JUN | JUL | AUG | SEP | OCT | NOV | DEC

GREY PARTRIDGE

Perdix perdix

Widespread from British Isles across NW Europe to S Sweden, S Finland and Baltic states into Russia. Mainly resident.

Distinctive round gamebird. Orange-brown head, grey neck and breast, and chestnut-barred flanks; brown upperparts and wings marked chestnut. Rufous outer tail. Blue-grey bill and legs.

When flushed, flies low over a short distance on whirring wings. Ground-nester in good cover. Both adults tend the chicks, which run easily on 1st day. Pair in family until Aug; small flocks (coveys) from then until *c.* Feb.

FACT FILE

LENGTH 29–31cm
WINGSPAN 45–48cm
WEIGHT 310–420g
FOOD Chiefly plant material; insects for chicks
EGGS 10–20, olive-brown
HABITAT Agricultural land with rough cover; also heaths and sand-dunes nearby
VOICE Call: like a rusty gate

JAN | FEB | MAR | APR | MAY | JUN | JUL | AUG | SEP | OCT | NOV | DEC

PHEASANT

Phasianus colchicus

Common and widespread, but only S Norway and S Sweden in Scandinavia. Native of Turkey and the Black Sea eastwards; introduced to our region by the Romans.

MALE Dark, metallic green head with red wattles; some have a white neck-ring. Body usually chestnut, covered with black chevrons; tail (over 35cm long) barred black. **FEMALE** Buffish brown, covered with dark marks, especially on upperparts.

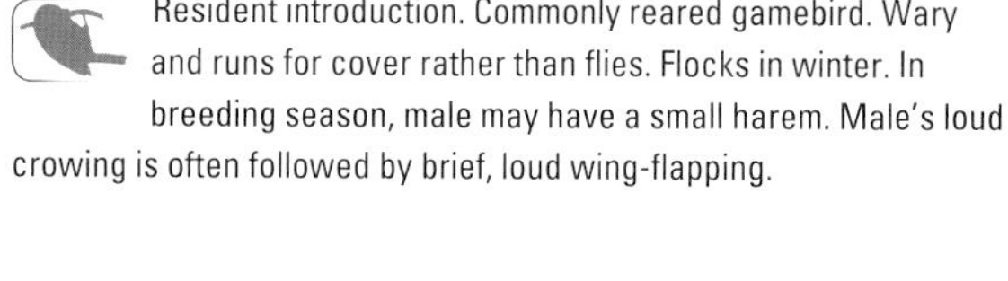

Resident introduction. Commonly reared gamebird. Wary and runs for cover rather than flies. Flocks in winter. In breeding season, male may have a small harem. Male's loud crowing is often followed by brief, loud wing-flapping.

RIGHT: *FEMALE*
BELOW: *MALE*

FACT FILE

LENGTH 53–89cm (male larger) WINGSPAN 70–90cm WEIGHT 720–1,800g
FOOD Omnivorous EGGS 8–15, olive-brown HABITAT Farmland, parkland, plantations
VOICE Harsh *kuttUk-kuttUK-kuttUK* when flushed. Males crow loudly in spring

ROCK DOVE

Columba livia

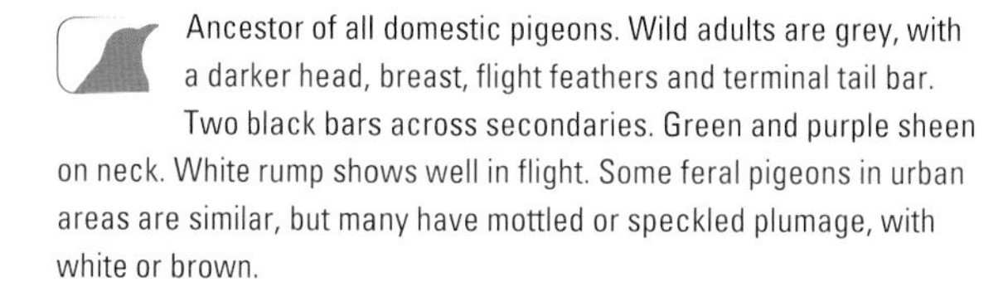

Ancestor of all domestic pigeons. Wild adults are grey, with a darker head, breast, flight feathers and terminal tail bar. Two black bars across secondaries. Green and purple sheen on neck. White rump shows well in flight. Some feral pigeons in urban areas are similar, but many have mottled or speckled plumage, with white or brown.

Widespread and common. Very hard to know the original distribution, because feral birds are so common.

Nests in colonies (hence the ease with which it was tamed). The best places to find birds in wild plumage in British Isles are far N Scotland and W Ireland.

FACT FILE

LENGTH 31–34cm WINGSPAN 63–70cm WEIGHT 238–370g FOOD Seeds, green leaves, insects EGGS 2, white HABITAT Traditionally, cliffs and rocky coastal landscapes or far inland VOICE *Oor-roo-coo*

JAN | FEB | MAR | APR | MAY | JUN | JUL | AUG | SEP | OCT | NOV | DEC

STOCK DOVE

Columba oenas

Widespread in British Isles and rest of region (not Iceland), mostly below 60°N. Largely resident in the W; birds from Germany eastwards winter in milder W and southwards to Iberia.

Mostly blue-grey. Broad blackish tips and trailing edges to wings show above and below; 2 narrow black bars on greater coverts; black band to end of tail. Glossy green sides to neck; breast vinaceous pink. Legs bright pink, bill horn with white cere.

Almost everywhere where trees are close to open ground for feeding. Nests in a tree-hole, rock crevice, ruined building or nestbox.

FACT FILE

LENGTH 32–34cm
WINGSPAN 63–69cm
WEIGHT 240–370g
FOOD Plant material, especially seeds
EGGS 2, white
HABITAT Copses, ancient parkland, city parks
VOICE *OOoo-er*, repeated quickly up to 10 times

JAN | FEB | MAR | APR | MAY | JUN | JUL | AUG | SEP | OCT | NOV | DEC

WOODPIGEON

Columba palumbus

Region's largest pigeon. Grey head and upperparts; green and purple sheen on neck; pink breast. **ADULT** Distinguished by white patch on each side of neck; unique white bar across (not along) wing coverts; grey tail with black terminal band above and grey, white and black below. **JUVENILE** Lacks neck patches.

Increasingly seen in parks and cities. Needs fresh water for drinking, as do other pigeons. Distinctive, deeply undulating display flight, with wing claps at top of each rise.

Widespread throughout except Iceland, mostly below 65°N. Largely resident in the W; populations increasingly migratory to the E and N. Can form very large flocks.

FACT FILE

LENGTH 40–42cm WINGSPAN 75–80cm WEIGHT 460–570g FOOD Weed and crop seeds EGGS 2, white HABITAT Wooded country bordered by fields VOICE *Coo COOO coo coo-coo-cuk*

JAN FEB MAR APR MAY JUN JUL AUG SEP OCT NOV DEC

COLLARED DOVE

Streptopelia decaocto

Throughout region, mostly S of *c.* 60°N, but spreading N. Mostly resident; juveniles disperse.

Dainty dove. **ADULT** Mostly pale pinkish buff with dusky-brown primaries. Upper tail brown with white tip, especially broad at corners; white with black base below, very noticeable in flight. Thin bar at sides of neck. **JUVENILE** Lacks neck bar.

Amazing spread from Balkans in 1930s; first bred in Sweden by 1951, and in SE England by 1955. In display flight, rises steeply then glides down, often calling a harsh *kurr* as it lands. Sings repeatedly from open perch.

FACT FILE

LENGTH 30–33cm WINGSPAN 47–55cm WEIGHT 115–240g FOOD Grain and weed seeds EGGS 2, white HABITAT Near human habitation, especially farms, and gardens with bird tables VOICE *Coo-OO-cuk*

CUCKOO

Cuculus canorus

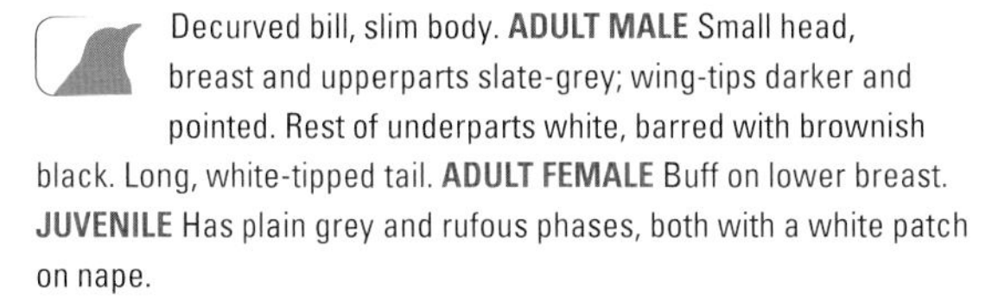

Decurved bill, slim body. **ADULT MALE** Small head, breast and upperparts slate-grey; wing-tips darker and pointed. Rest of underparts white, barred with brownish black. Long, white-tipped tail. **ADULT FEMALE** Buff on lower breast. **JUVENILE** Has plain grey and rufous phases, both with a white patch on nape.

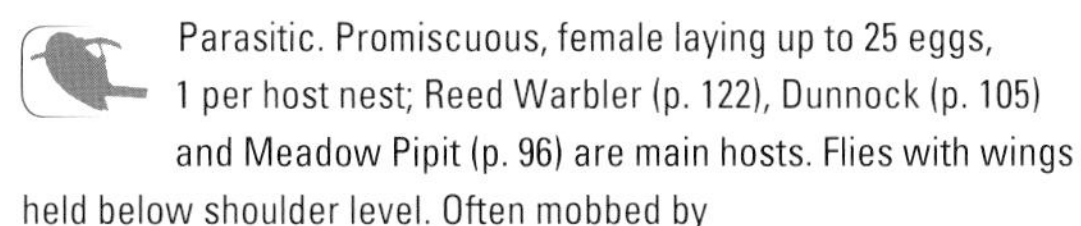

Parasitic. Promiscuous, female laying up to 25 eggs, 1 per host nest; Reed Warbler (p. 122), Dunnock (p. 105) and Meadow Pipit (p. 96) are main hosts. Flies with wings held below shoulder level. Often mobbed by small birds.

Summer visitor throughout region, but not Iceland or open tundra. Winters in Africa, mostly S of Equator.

RIGHT: *JUVENILE*
BELOW: *ADULTS*

cf. ***Sparrowhawk***

barred underparts

FACT FILE

LENGTH 32–34cm WINGSPAN 55–60cm WEIGHT 100–160g FOOD Beetles, hairy caterpillars EGGS Very variable, resembling host's eggs HABITAT *See* hosts' details VOICE Male's call: famous *cuc-oo*. Female's call: bubbling

JAN FEB MAR **APR MAY JUN JUL AUG SEP** OCT NOV DEC

OSPREY

Pandion haliaetus

In flight, wings are often seen as a gull-like 'M'. Long-winged, with a white head and underparts, and dark brown upperparts and upperwing.

Fish are caught in talons after a shallow dive; gets into trouble at fish farms. Large stick-nest on a treetop or specially erected pole, often used year after year.

Persecuted to extinction in much of Europe by the early 20th century. Naturally recolonized in UK in 1954, now c. 300 pairs; widespread from Scandinavia eastwards. Seen more widely in autumn, usually singly, on migration to sub-Saharan Africa.

FACT FILE

LENGTH 52–60cm
WINGSPAN 152–167cm
WEIGHT 1,120–2,050g (female larger)
FOOD Fish
EGGS 2–3, creamy white with red-brown blotches and pale grey underneath
HABITAT Year-round on lakes, rivers, coasts
VOICE High-pitched whistling and yelping in breeding season

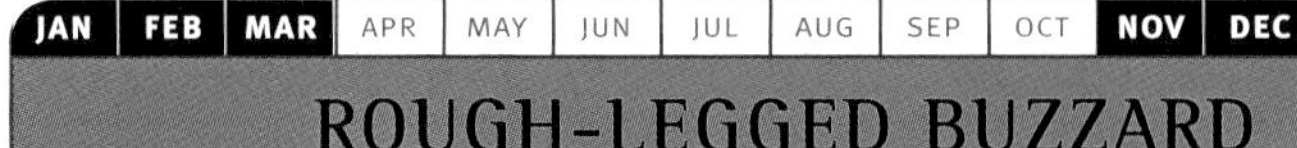

ROUGH-LEGGED BUZZARD

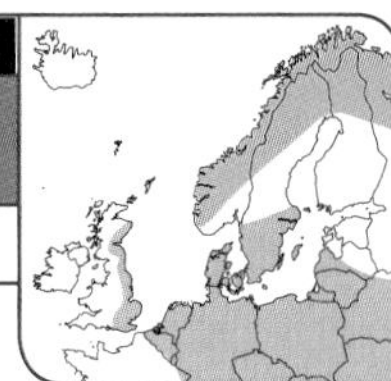

Buteo lagopus

Closely resembles Buzzard (below). Plumage variable. Whitish head and underparts, dark belly, largely white underwing with black tips and carpal patch. Tail pattern diagnostic.

Nests mostly on rocky ledges or sheltered sites on ground. Numbers and distribution vary in relation to abundance or scarcity of small mammalian prey, especially lemmings. Often hovers.

Summer visitor to Scandinavia and N Russia. Winters in S Sweden and Denmark southwards; a few occur each year in E Britain.

FACT FILE

LENGTH 50–60cm WINGSPAN 120–150cm WEIGHT 600–1,660g (female larger) FOOD Chiefly small mammals EGGS 3–4; white, blotched reddish brown HABITAT Breeds in low-lying, treeless tundra; winters in open country VOICE Cat-like *mee-oo*

JAN FEB MAR APR MAY JUN JUL AUG SEP OCT NOV DEC

BUZZARD

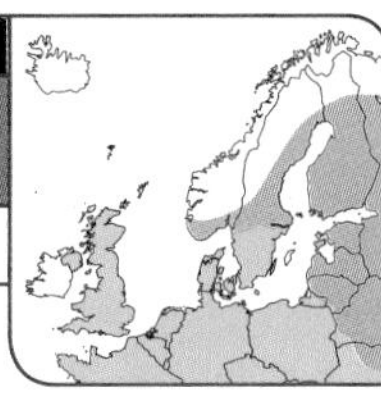

Buteo buteo

Upperparts dark brown; underparts white with very variable brown streaks and bars, most noticeable across breast. Pale underwings with dark carpal patch. Glides with wings in noticeable 'V'.

Most often seen spiralling over its territory. Often hunts by waiting for prey passing below its perch. Bulky tree nest built of sticks with a soft lining.

Across the region; rarely in Eire, not in Iceland or N Fennoscandia; increasing in U.K. Most W birds are resident; E birds winter S and W.

FACT FILE

LENGTH 51–57cm WINGSPAN 113–128cm WEIGHT 500–1,300g FOOD Mostly small rodents EGGS 2–4; white, with variable red or brown markings HABITAT Farmland, forest clearings, moorland edges with mature trees VOICE Mewing, far-carrying *pee-oo, pee-oo*

JAN | FEB | MAR | APR | MAY | JUN | JUL | AUG | SEP | OCT | NOV | DEC

SPARROWHAWK

Accipiter nisus

Found throughout region except Iceland and the far N. Birds breeding in N Scandinavia migrate to winter within rest of range; others are resident or disperse locally.

Flies with a rapid burst of wingbeats, then a glide, revealing barred underwing and tail. **MALE** Grey above; underparts pale, narrowly barred orange. **FEMALE** Brown above, barred brown below. **JUVENILE** Brownish upperparts and pale underparts with broad, brown barring.

Hunts by dashing from perch to perch, snatching prey in flight or off its perch – including bird tables! Male takes small songbirds; female can kill a dove. Quite often seen being mobbed by Swallows (p. 94) or Starlings (p. 143) above hunting ground.

RIGHT: *MALE*
BELOW: *JUVENILE*

FACT FILE

LENGTH 28–38cm WINGSPAN 55–70cm WEIGHT Male, 144g; female, 264g FOOD Almost entirely birds EGGS 4–6; bluish white, streaked brown or reddish brown HABITAT Woods, farmland, gardens VOICE Shrill, chattering *gek*

JAN FEB MAR APR MAY JUN JUL AUG SEP OCT NOV DEC

KESTREL

Falco tinnunculus

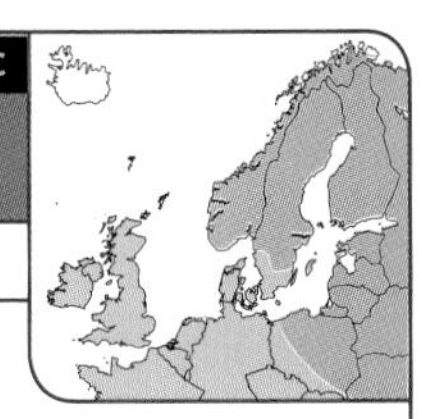

Throughout region except Iceland. Mostly resident, except in the far E and N, whose birds migrate to within rest of region. Commonest daytime raptor in many places.

MALE Chestnut upperparts with blackish spots; dark-streaked buff underparts; bluish-grey head, rump and tail. **FEMALE** Reddish brown above with dark barring, paler below with dark streaks, and barred brown tail. Pointed wings.

Habitually hovers looking for prey, suddenly dropping onto it. Nests in old crows' nests, cliff ledges, hollow trees, large nestboxes and ruined buildings.

TOP: *FEMALE;* ABOVE: *MALE*

FACT FILE

LENGTH 32–35cm WINGSPAN 71–80cm WEIGHT 136–314g (female heavier)
FOOD Especially voles; in warmer times, insects EGGS 3–6; white, heavily marked red-brown
HABITAT Farmland, hill country, moorland, rocky coasts VOICE Clear *kee-kee-kee*

JAN FEB MAR APR **MAY** **JUN** **JUL** **AUG** SEP OCT NOV DEC

HOBBY

Falco subbuteo

Long-distance summer migrant to S Britain and eastwards to around Baltic. Thinly distributed in the W. Winters in S third of Africa in wet season, when insects are abundant.

Comparatively short tail and long, scimitar-shaped wings – can look like a large Swift (p. 89). Dark slate-grey above, heavily streaked below; patterned head of dark cap, whitish cheeks and throat.

Pursues prey in flight. Nests in old crows' nests or similar. Young are fed mainly on birds, especially species in which the male specialises – hirundines, larks and pipits.

FACT FILE

LENGTH 30–36cm WINGSPAN 82–92cm WEIGHT 131–340g (female heavier)
FOOD Almost exclusively aerial prey – insects and birds EGGS 3; brownish yellow, finely spotted red-brown HABITAT Open lowland with clumps of trees VOICE Rapid *kew kew kew kew*

JAN | FEB | MAR | APR | MAY | JUN | JUL | AUG | SEP | OCT | NOV | DEC

PEREGRINE

Falco peregrinus

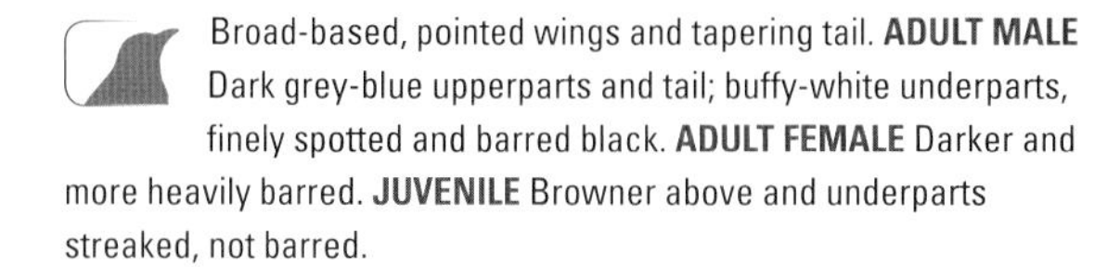

Broad-based, pointed wings and tapering tail. **ADULT MALE** Dark grey-blue upperparts and tail; buffy-white underparts, finely spotted and barred black. **ADULT FEMALE** Darker and more heavily barred. **JUVENILE** Browner above and underparts streaked, not barred.

Nests on cliff ledges and city towers. Hunts by circling high, spotting prey below, then diving at incredible speed for an aerial kill.

Not Iceland; from British Isles eastwards in suitable habitats. Suffered marked declines in the mid-20th century due to poisons in the food chain; has since recovered well in British Isles.

ABOVE: *JUVENILE*
LEFT: *ADULT*

FACT FILE

LENGTH 36–48cm WINGSPAN 95–110cm WEIGHT 300–580g (female 15% larger) FOOD Birds, especially seabirds and pigeons EGGS 3–4, buff or cream with heavy red-brown markings HABITAT Chiefly open country with coastal or inland cliffs, including quarries VOICE Shrill *kek-kek-kek-kek*

JAN | FEB | MAR | APR | MAY | JUN | JUL | AUG | SEP | OCT | NOV | DEC

TAWNY OWL

Strix aluco

Widespread resident, but not Iceland, Faeroes, Scottish islands or Ireland, and mostly in S in Scandinavia.

Large, round head; broad, rounded wings. Greyish-brown face mask. Most of plumage rufous brown in W birds, greyer in E birds, streaked above and below; wings have 2 broken white bars.

Nests in tree-holes, old nests and special nestboxes. Nocturnal. Roosts in thick cover. Mobbed if found by small birds. Hunts at night on silent wings. Indigestible bones and fur are ejected in a pellet.

FACT FILE

LENGTH 37–39cm
WINGSPAN 94–104cm
WEIGHT 340–650g
FOOD Mainly rodents, also birds, insects
EGGS 2–5, white
HABITAT Woods, copses, parkland, even town suburbs
VOICE Male's call: *hoo-hoo*. Both birds: *ke-wick*. Together, calls form classic *tu-whit-to-woo*

JAN | FEB | MAR | APR | MAY | JUN | JUL | AUG | SEP | OCT | NOV | DEC

SHORT-EARED OWL

Asio flammeus

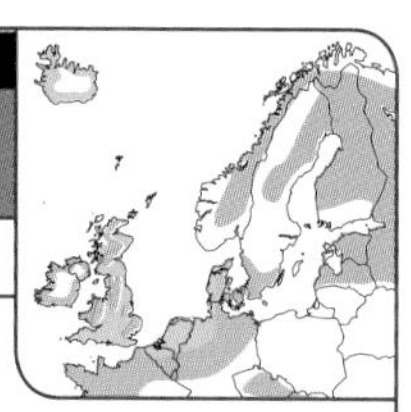

Pale face mask, short ear tufts (usually invisible), eyes outlined in black and white. Streaked breast but whiter belly; upperparts golden brown, heavily marked with black. Wing has big yellowy-buff primary patch; wing-tip looks as if tipped black, and trailing edge of wing is narrowly white.

Nests on ground in optimal habitat. Hunts at all times of day and night, quartering the ground.

Breeds in Iceland, N Britain, Low Countries, N Germany, Denmark, Sweden, Norway, Finland and N Baltic states eastwards. N and E populations migrate S and W.

FACT FILE

LENGTH 37–39cm WINGSPAN 95–110cm WEIGHT 260–425g FOOD Voles especially EGGS 4–8, white HABITAT Tundra, moorland, young plantations VOICE Low *hoo hoo hoo*

JAN | FEB | MAR | APR | MAY | JUN | JUL | AUG | SEP | OCT | NOV | DEC

BARN OWL

Tyto alba

Widespread across region, but no further N than Scotland, Denmark and Latvia. Sedentary.

British and French birds have mottled grey and buff upperparts and wing coverts, and pure white underparts; eastern birds have mainly grey upperparts and yellowish-orange underparts, finely speckled with brown. In flight, underwing is white, and flight feathers are buff with faint darker bars. Heart-shaped facial mask white, thinly outlined in dark brown; eyes black.

Crepuscular and nocturnal. Decreasing owing to loss of nest sites in tree-holes, buildings and barns. Recovers where nestboxes are provided. At dusk, looks ghostly while quartering and hovering over open ground.

TOP AND ABOVE: *WESTERN BIRD*

FACT FILE

LENGTH 33–35cm WINGSPAN 85–93cm WEIGHT 240–360g FOOD Small mammals EGGS 4–7, white HABITAT Farmland, rough grassland, hedges, copses VOICE Drawn-out screech, *shreeee*

JAN | FEB | **MAR** | **APR** | **MAY** | **JUN** | **JUL** | **AUG** | **SEP** | OCT | NOV | DEC

NIGHTJAR

Caprimulgus europaeus

At dusk, appears to be dusky brown, except that male has white spots near base of outer primaries and outer tips of tail; both sexes have white throat. Plumage is actually cryptic: an intricate pattern of blotches, streaks and bars, of silvery grey, black, buff and rufous. Bill is small, gape large.

Crepuscular, aerial feeder; silent flight on long wings with twists and turns, and hovering.

Summer visitor to British Isles (not N Scotland), France, eastwards to S Scandinavia, Baltic states and Russia. Winters in sub-Saharan Africa away from forest.

LEFT: *FEMALE*
BELOW LEFT: *MALE*

FACT FILE

LENGTH 26–28cm
WINGSPAN 57–64cm
WEIGHT 65–100g
FOOD Insects
EGGS 2; cream, marked with cream and brown
HABITAT Open woodland, moorland, new plantations, heath, sand-dunes
VOICE Male's song: like breathing a purring *rrrrrrrrrrrr*, sometimes for several minutes. Both sexes: *coo-ik*

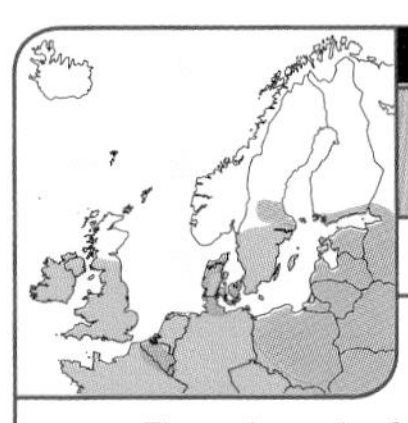

JAN FEB MAR APR MAY JUN JUL AUG SEP OCT NOV DEC

KINGFISHER

Alcedo atthis

Through much of region except N Scotland and most areas N of 60°N. In winter, birds in Finland and from Russia to Poland migrate W to ice-free waters.

Unmistakable: bright blue upperparts and orange underparts. Tone of blue changes with the light – back and rump can flash with an intense sheen. Head pattern distinctive: crown, moustache and nape blue; lores and ear coverts orange; throat and patch behind ear coverts white. Bill all black in male, much of lower mandible reddish in female.

Territorial in breeding season. Digs a tunnel with an egg chamber at end, usually 45–90cm long, in a bank generally overlooking water.

FACT FILE

LENGTH 16–17cm WINGSPAN 24–26cm WEIGHT 36–46g FOOD Mainly fish EGGS 6–7, white HABITAT Still or gently flowing fresh waters VOICE Shrill *ti-tee*

SWIFT

Apus apus

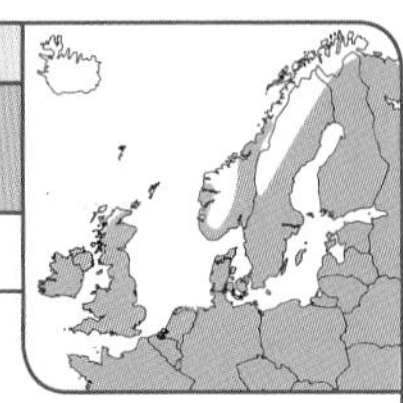

Throughout much of region, but few in far N Scotland, and not on N islands, Iceland or much of far N Scandinavia. Winters S of the Sahara.

All blackish brown with a whitish throat. Long, narrow wings. Short, forked tail. Sometimes misidentified as a Swallow (p. 94), but that has white underparts.

Natural nesting habitat has been almost wholly replaced by buildings. Wanders widely for food – anywhere it can hunt insects in the air – so seen over a wide variety of habitats. Fast, agile flight, often seeming to have alternate wingbeats. Very noticeable at nest site, where several pairs may race around screaming.

FACT FILE

LENGTH 16–17cm WINGSPAN 42–48cm WEIGHT 31–56g FOOD Insects EGGS 2–3, white HABITAT Aerial. Nests in crevices in cliffs, crags, buildings VOICE Shrill *sreeeeee*

Not Ireland, N Scotland or Iceland. Otherwise widespread eastwards, and N to 60°N and slightly beyond in Norway and Sweden.

JAN	FEB	MAR	APR	MAY	JUN	JUL	AUG	SEP	OCT	NOV	DEC

GREEN WOODPECKER

Picus viridis

ADULT Yellowish green above and pale greenish grey below, with lightly barred flanks; in flight, shows bright yellow rump and cream bars across dark brown primaries. Distinctive head pattern: red crown and nape, black around eye, and black moustache, male's with a red centre. **JUVENILE** White-spotted upperparts, brownish-black bars on face and underparts.

Very specialised, feeding mostly on the ground, digging into ant-nests and extending its very long, sticky tongue. Drums rarely.

RIGHT: *FEMALE AT NEST HOLE*
BELOW: *JUVENILE*

FACT FILE

LENGTH 31–33cm WINGSPAN 40–42cm WEIGHT 170–250g FOOD Almost exclusively ants and their pupae EGGS 5–7, white HABITAT Deciduous woodland, orchards, parkland VOICE Laughing *cue-cue-cue*

JAN | FEB | MAR | APR | MAY | JUN | JUL | AUG | SEP | OCT | NOV | DEC

GREAT SPOTTED WOODPECKER

Dendrocopus major

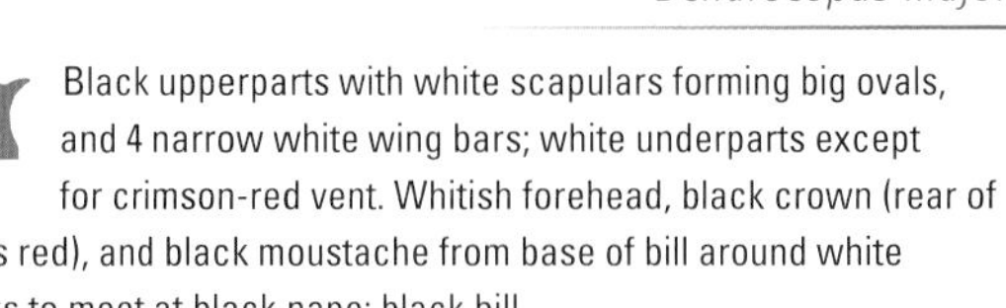

Black upperparts with white scapulars forming big ovals, and 4 narrow white wing bars; white underparts except for crimson-red vent. Whitish forehead, black crown (rear of male's red), and black moustache from base of bill around white cheeks to meet at black nape; black bill.

Eats many conifer seeds in winter; if seed source fails, many erupt S and W. Uses stiff tail to help control climb up tree; has bouncing flight. In spring, drums on a dead branch with bill for *c.* 0.5 seconds, sound dying away at end.

Resident throughout region except Ireland, Northern Isles, Iceland and far N Scandinavia.

ABOVE: *MALE;* RIGHT: *JUVENILE*

FACT FILE

LENGTH 22–23cm WINGSPAN 34–39cm WEIGHT 70–100g
FOOD Mainly insects and larvae EGGS 4–7, white
HABITAT Woodland VOICE Sharp *kik*

JAN | FEB | MAR | APR | MAY | JUN | JUL | AUG | SEP | OCT | NOV | DEC

SKYLARK

Alauda arvensis

Widespread through much of region except Iceland, high Arctic and spine of Norway. Mostly resident; N and E populations migrate to winter in milder W. Widespread decline in W owing to agricultural intensification.

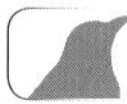

Streaky grey-brown bird, which walks rather than hops. Streaks on breast sharply divided from rest of white underparts. In flight, shows white trailing edge to broad wings, and white outer tail feathers.

Rarely perches on trees. Most often noticed when male flies up steeply, singing very varied, melodious, non-stop outpouring at 'song post' in the sky, *c.* 100m up, for several minutes, followed by slow descent.

FACT FILE

LENGTH 18–19cm WINGSPAN 30–36cm WEIGHT 32–55g FOOD Insects, seeds EGGS 3–5; greyish, spotted brown HABITAT Farmland, heaths, moors, coastal dunes, even airfields VOICE Liquid chirrup; *see also* 'Habits'

JAN | FEB | **MAR** | **APR** | **MAY** | **JUN** | **JUL** | **AUG** | **SEP** | **OCT** | NOV | DEC

SAND MARTIN

Riparia riparia

Summer visitor throughout region except Iceland and high Arctic. Winters in sub-Saharan Africa.

Smaller than the House Martin (p. 95) and Swallow (p. 94). Brown upperparts; white underparts with broad, well-marked brown breast-band. Forked tail. Dusky underwing.

Very closely attached to water. Nests colonially (usually <50 pairs) in tunnels in sandy or earthy banks, mostly 45–90cm long, dug by both birds in a pair. Builds nest of grass, leaves and feathers at end (unlike bare chamber of Kingfisher, p. 88). Aerial feeder, chiefly over water. Gregarious on migration, roosting in hundreds or thousands in reedbeds.

FACT FILE

LENGTH 12cm WINGSPAN 27–29cm WEIGHT 10–19g FOOD Airborne insects EGGS 4–6, white HABITAT Riverbanks, gravel pits, sandy sea cliffs VOICE Harsh *twitter*

JAN | FEB | MAR | **APR** | **MAY** | **JUN** | **JUL** | **AUG** | **SEP** | **OCT** | NOV | DEC

SWALLOW

Hirundo rustica

Summer visitor throughout, except Iceland and high Arctic. Winters in Africa, mostly S of Equator.

Deeply forked tail with long, thin outer streamers (2–7cm, male's longer than female's or juvenile's). **ADULT** Upperparts and breast-band shiny blue-black, flight feathers dull black, underparts off-white. Deep red forehead, chin and throat. **JUVENILE** Red is replaced by reddish buff.

Traditional nest site rare, in cave entrance; now mostly on a ledge in a farm building, shed, stable or church porch. Gregarious on migration, gathering on wires and roosting in reedbeds. Feeds on the wing.

RIGHT: *ADULT AND YOUNG*
BELOW: *MALE*

FACT FILE

LENGTH 17–19cm
WINGSPAN 32–35cm
WEIGHT 14–24g
FOOD Flying insects
EGGS 4–5; white, lightly spotted red-brown
HABITAT Open country, often near water, for aerial feeding
VOICE Sharp *witt* or *witt-witt*. Warbling song, on the wing

JAN FEB MAR **APR MAY JUN JUL AUG SEP OCT** NOV DEC

HOUSE MARTIN

Delichon urbica

Upperparts deep metallic blue; flight feathers and tail dull black, in sharp contrast to pure white underparts and rump. Deeply forked tail. Legs fully feathered white; pink feet.

Colonial nester (mostly <5 pairs). Few birds still nest on traditional site under a cliff's rocky overhang; most now under eaves of buildings. Builds a cup-shaped mud nest, lined with grass and feathers, up against overhang, with entrance at top. Migrants roost in trees, and maybe high in the air then and during the breeding season.

Summer visitor, except Iceland and high Arctic. Winters across sub-Saharan Africa.

FACT FILE

LENGTH 12.5cm WINGSPAN 26–29cm WEIGHT 15–23g FOOD Flying insects EGGS 3–5, white HABITAT Aerial over cliffs, villages, farms VOICE Hard *chirrrp* and twittering

Found throughout region. N birds migrate to winter in British Isles, from Denmark to France, and S to N Africa.

JAN | FEB | MAR | APR | MAY | JUN | JUL | AUG | SEP | OCT | NOV | DEC

MEADOW PIPIT

Anthus pratensis

Distinguishable from Tree Pipit (p. 97) by voice. Grey-brown above with olive tint; dark brown stripes on crown and back. Centres of coverts and tertials blackish, edges pale, wing bars dull. Underparts white with buff at sides; breast and flanks streaked. White outer tail feathers. Base of bill yellow.

Deserts uplands in winter for farmland and seashore. Sings in flight: weak notes, gathering speed as bird rises, and ending in a trill as it 'parachutes' down. One of the main hosts of the Cuckoo (p. 77).

FACT FILE

LENGTH 15cm WINGSPAN 22–25cm WEIGHT 14–24g FOOD Mainly invertebrates EGGS 3–5; grey to reddish, well marked with blackish HABITAT Heaths, rough grassland, moors, sand-dunes VOICE Call: weak *tsip tsip*. Flight song: *see* 'Habits'

JAN | FEB | MAR | APR | **MAY** | **JUN** | **JUL** | **AUG** | **SEP** | OCT | NOV | DEC

TREE PIPIT

Anthus trivialis

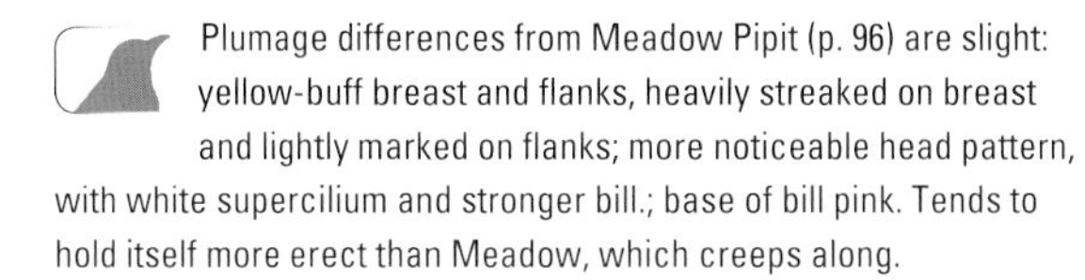

Plumage differences from Meadow Pipit (p. 96) are slight: yellow-buff breast and flanks, heavily streaked on breast and lightly marked on flanks; more noticeable head pattern, with white supercilium and stronger bill.; base of bill pink. Tends to hold itself more erect than Meadow, which creeps along.

The only pipit that needs scattered trees for song posts. Song usually starts from a treetop; male then flies up trilling, and finishes with a musical *SEE-a-SEE-a-SEE-a* as it glides back to the same or a nearby perch.

Summer visitor to region except N islands, Iceland and Ireland. Winters in tropical Africa.

FACT FILE

LENGTH 15cm WINGSPAN 25–27cm WEIGHT 18–30g FOOD Mainly insects EGGS 2–6, extremely variable ground colour and markings HABITAT Rough agricultural land, heaths, young plantations, birch scrub VOICE Hoarse *teez*. Song: *see* 'Habits'

Breeds on coasts of NW France, British Isles and Scandinavia. N birds disperse to winter on coasts of S British Isles, and from France to Denmark.

JAN | FEB | MAR | APR | MAY | JUN | JUL | AUG | SEP | OCT | NOV | DEC

ROCK PIPIT

Anthus petrosus

Predominantly dark and streaked. Upperparts brownish grey with an olive tint, mantle streaked darker, wing feathers with darker centres; wing bars indistinct buff. Underparts dirty white, tinged buff on breast and flanks, both of these heavily streaked. Outer tail feathers greyish white.

Nests in a hollow in a cliff or bank by the shore, or under thick vegetation above high-tide line. Sings in flight.

FACT FILE

LENGTH 16–17cm WINGSPAN 23–28cm WEIGHT 21–32g FOOD Invertebrates EGGS 4–6; whitish, heavily spotted olive-brown HABITAT Rocky islands, sea cliffs, rocky seashores VOICE Short, sharp *pseep*. Song: accelerating series of notes with final trill

JAN | FEB | MAR | APR | MAY | JUN | JUL | AUG | SEP | OCT | NOV | DEC

PIED WAGTAIL/WHITE WAGTAIL

Motacilla alba yarrelli/Motacilla alba alba

PIED (above) breeds in British Isles and nearby Continental coast; afterwards, birds are short-distance migrants. WHITE (below) is found everywhere else; migrates to winter in W Europe and Mediterranean basin.

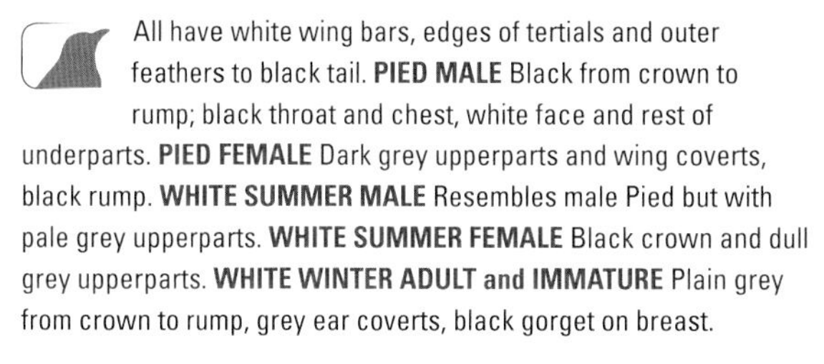

All have white wing bars, edges of tertials and outer feathers to black tail. **PIED MALE** Black from crown to rump; black throat and chest, white face and rest of underparts. **PIED FEMALE** Dark grey upperparts and wing coverts, black rump. **WHITE SUMMER MALE** Resembles male Pied but with pale grey upperparts. **WHITE SUMMER FEMALE** Black crown and dull grey upperparts. **WHITE WINTER ADULT and IMMATURE** Plain grey from crown to rump, grey ear coverts, black gorget on breast.

Ground feeders, walking with a nodding head and wagging tail. Territorial breeders, but communal afterwards.

LEFT: *FEMALE PIED WAGTAIL*
BELOW: *MALE WHITE WAGTAIL*

FACT FILE

LENGTH Both 18cm
WINGSPAN 25–30cm
WEIGHT 17–27g
FOOD Insects
EGGS 5–6; greyish white, finely speckled grey-brown
HABITAT Especially around farms and cultivation with water nearby
VOICE Call: *tschizzick*. Song: lively twittering

JAN	FEB	MAR	**APR**	**MAY**	**JUN**	**JUL**	**AUG**	**SEP**	OCT	NOV	DEC

YELLOW WAGTAIL

Motacilla flava

All summer visitors, wintering in tropical Africa; (1) breeds in central Europe (purple on map), (2) breeds in N Fennoscandia (yellow), (3) breeds in British Isles and coastal France (orange). Other identifiable subspecies from the S and E occur on migration through the region.

Three subspecies in region: (1) Blue-headed (*M. f. flava*); (2) Grey-headed (*M. f. thunbergi*); Yellow (*M. f. flavissima*). **MALE** All have yellowish-green back and rump; wing coverts black with 2 pale greenish-yellow wing bars; tail black with white outer feathers; whole of underparts bright yellow. (1) Bluish-grey crown and darker ear coverts, and white supercilium. (2) Dark slate-grey crown and nape, and almost black cheeks, with or without whole or faint white supercilium. (3) Yellow forehead, yellowish-green crown and cheeks, yellow supercilium, and upperparts washed with yellow. **FEMALE** All are duller and less yellow; form best identified when with male.

All territorial ground nesters. Gregarious on migration. Often hunt insects around the feet of cattle. Most often spotted when a passing bird calls overhead.

LEFT: *MALE YELLOW WAGTAIL;* ABOVE RIGHT: *MALE BLUE-HEADED WAGTAIL;* BELOW RIGHT: *MALE GREY-HEADED WAGTAIL*

FACT FILE

LENGTH 17cm WINGSPAN 23–27cm WEIGHT 15–20g FOOD Small invertebrates EGGS 4–6; whitish, densely spotted brown HABITAT (1) and (3) in lowland water meadows, riversides, fens; (2) in *Sphagnum* and peat bogs with shrubs or stunted trees VOICE Shrill, rather plaintive *tsweep*

JAN | FEB | MAR | APR | MAY | JUN | JUL | AUG | SEP | OCT | NOV | DEC

GREY WAGTAIL

Motacilla cinerea

British Isles and France, eastwards to S Norway and Sweden. N birds especially migrate to winter in rest of breeding range.

Constantly wags tail. **ADULT MALE** Upperparts blue-grey, rump yellow-green. White supercilium, white lower border to cheeks, black bib. Underparts yellow, brightest under tail. Wings and tail black, with white wing bar and outer tail. Legs brownish pink, not dark as in other wagtails. **ADULT FEMALE** Similar but bib white or mottled black, and underparts paler. **IMMATURE** White throat, and yellow only under tail (beware confusion with Yellow Wagtail, p. 100).

Often seen flycatching. Constantly wags its tail.

TOP: *WINTER FEMALE*
LEFT: *MALE*

FACT FILE

LENGTH 18–19cm WINGSPAN 25–27cm WEIGHT 14–22g FOOD Mainly insects EGGS 4–6; whitish, faintly marked grey HABITAT Swift-flowing, rocky, mostly shallow rivers and streams with wooded banks. In winter, also by lowland lakes and streams VOICE Distinctive metallic *tzi-tzi*

Iceland, British Isles and across region except N Fennoscandia. Mainly resident; most N and E populations migrate S and W.

JAN	FEB	MAR	APR	MAY	JUN	JUL	AUG	SEP	OCT	NOV	DEC

WREN

Troglodytes troglodytes

Unmistakable tiny bird. Reddish brown above, whitish brown below, all with fine, darker barring.

Active, hunting invertebrates in low cover. Rapid, straight, whirring flight between bushes. Male builds a completely domed nest; female lines it. Song amazingly loud for a small bird.

FACT FILE

LENGTH 9–10cm WINGSPAN 13–17cm WEIGHT 7–12g FOOD Insects, spiders
EGGS 5–8, white with some dark speckles at broad end HABITAT Dense undergrowth: woodland, islands, gardens, hedgerows, moorland with heather and Bracken
VOICE Rattling alarm and hard *tic tic tic*. Song: notes and trills for 4–6 seconds

JAN | FEB | MAR | APR | MAY | JUN | JUL | AUG | SEP | OCT | NOV | DEC

DIPPER

Cinclus cinclus

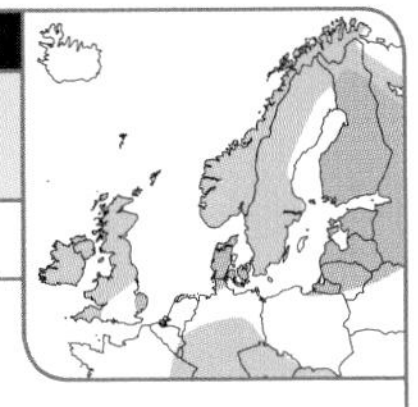

N and W British Isles; some in suitable habitats in France and to Scandinavia eastwards.

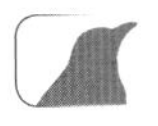

ADULT Head and nape chocolate-brown; rest of upperparts more slatey with black feather edges. Underparts white from chin to breast, the rest blackish brown (birds in British Isles have a chestnut band between white and brown). Noticeable white eyelid when bird blinks. **JUVENILE** Mottled grey. **FIRST-YEAR** White tips to wing coverts.

Territorial. Flight low and direct along river. Bobs on hinged legs on boulder. Often feeds underwater, walking along bottom, or swims with flicking wings.

RIGHT: *JUVENILE*
BELOW: *BRITISH ADULT*

FACT FILE

LENGTH 18cm WINGSPAN 25–30cm WEIGHT 50–72g FOOD Mainly aquatic invertebrates EGGS 4–5, white HABITAT Fast-flowing streams and rivers with rocks and boulders, in hilly regions VOICE Song: rippling warble. Call: short, sharp *zit zit* as it flies

Breeds in Arctic Sweden, Finland and Russia eastwards. Regularly winters in E and N Europe.

JAN | FEB | MAR | APR | MAY | JUN | JUL | AUG | SEP | OCT | NOV | DEC

WAXWING

Bombycilla garrulus

Unmistakable. Head, back and wing coverts vinaceous brown; rump grey, tail black with yellow tip; narrow black eye-stripe with chestnut above and below; black bib; reddish vent. Amazing wing pattern: black or blackish flight feathers; white-tipped primary coverts; white-tipped secondaries with wax-red tip beyond that; primaries' tips edged white and yellow.

In the breeding season, feeds by flycatching from treetops. If fruit crop fails, large winter flocks erupt to find trees and bushes covered with red berries, as far W as France and Ireland.

FACT FILE

LENGTH 18cm WINGSPAN 32–35cm WEIGHT 50–83g FOOD Mainly insects in summer, berries in winter EGGS 5–6; bluish, speckled with black HABITAT Conifer forests VOICE Twittering song. Call: ringing *sirrrrrrr*

JAN | FEB | MAR | APR | MAY | JUN | JUL | AUG | SEP | OCT | NOV | DEC

DUNNOCK

Prunella modularis

Warm brown above, and lead-grey head and below, recalling female House Sparrow (p. 144). Brown upperparts streaked with blackish brown; sides of breast and flanks buff-brown, streaked darker; brownish crown and ear coverts.

Unobtrusive ground feeder, creeping along, belly near ground, but often sings from an exposed perch. One of the main hosts of the Cuckoo (p. 77)

Throughout region except Iceland and high Arctic. N and Continental populations migrate S and W. Resident in British Isles and W France, or disperses short distances.

FACT FILE

LENGTH 15cm WINGSPAN 19–21cm WEIGHT 14–25g FOOD Insects, seeds EGGS 4–6, blue HABITAT Woodland along river valleys, hedgerows, spinneys, gardens, low scrub on moorland VOICE Call: loud *tseep*. Song: short, loud, high-pitched warble

JAN | FEB | MAR | APR | MAY | JUN | JUL | AUG | SEP | OCT | NOV | DEC

ROBIN

Erithacus rubecula

Throughout except Iceland and the far N. Largely resident in British Isles; birds from the N and E winter in the W and southwards to N Africa.

ADULT Upperparts, wings and tail olive-brown. Forehead, cheeks and breast orange, separated from brown by band of blue-grey. Flanks warm buff, rest of underparts white. **JUVENILE** Brown above, with buff spots; buff below, with small, darker spots on breast.

Britain's national bird. In British Isles and the W, is a regular in parks, gardens and farmland. Will follow animals and humans to seek disturbed food.

LEFT: *JUVENILE;* ABOVE: *ADULT*

FACT FILE

LENGTH 14cm WINGSPAN 20–22cm WEIGHT 13–25g FOOD Insects, especially beetles HABITAT Woods and copses with open spaces on which to feed VOICE Call: scolding *tic tic* and high-pitched *tswee*. Song: melodious, sad-sounding warble

JAN FEB **MAR APR MAY JUN JUL AUG SEP OCT** NOV DEC

WHEATEAR

Oenanthe oenanthe

Found from the Arctic southwards, from sea-level to >3,000m. Summer visitor, wintering in Africa at 0–20°N. Often the first songbird to arrive in the region, commonly in Mar. Autumn migration is protracted, from Aug to Oct.

MALE Body plumage pattern is diagnostic and distinctive, with white tail tipped by a broad black upside-down 'T'. **FEMALE** Similarly patterned, but wings, crown, cheek-patch and back all brown-toned, and underparts usually buffer.

Nests in a hole in a rock or wall, or in a Rabbit burrow. Wary; if pressed, dashes away showing tail pattern.

TOP: *JUVENILE*
LEFT: *SUMMER MALE*

FACT FILE

LENGTH 14.5–15.5cm WINGSPAN 26–32cm WEIGHT 21–30g FOOD Mainly insects and other invertebrates; berries in autumn EGGS 5–6, pale blue HABITAT Tundra, sand-dunes, cliff tops, moors, mountains VOICE Call: *chak*. Alarm: *weet-chak, chak*. Song: energetic, short warble

JAN	FEB	MAR	**APR**	**MAY**	**JUN**	**JUL**	**AUG**	**SEP**	**OCT**	NOV	DEC

REDSTART

Phoenicurus phoenicurus

Summer visitor throughout except Iceland, Ireland, Northern Isles and high Arctic. Winters in tropical Africa N of Equator.

All have orange-chestnut rump and tail (with dark centre). **ADULT MALE** Forehead white; rest of upperparts blue-grey, including wing coverts; wings blackish brown; throat and upper breast black, sharply divided from orange underparts. **ADULT FEMALE and IMMATURE** Grey-brown upperparts and pale orange-buff underparts, fading to whitish belly.

Tail very noticeable in flight, and when perched, quivering up and down. Nests in a hole in a tree or wall, or a nestbox.

TOP: *FEMALE;* ABOVE: *MALE*

FACT FILE

LENGTH 14cm WINGSPAN 21–24cm WEIGHT 13–20g FOOD Mainly insects, some fruit EGGS 5–7, blue HABITAT Old deciduous woodland and hill country with scattered trees and stone walls VOICE Plaintive *wheet*, often linked with *tooick*. Brief, melodious song

JAN | FEB | MAR | APR | MAY | JUN | JUL | AUG | SEP | OCT | NOV | DEC

BLACK REDSTART

Phoenicurus ochruros

All have dark-centred orange tail like Redstart (p. 108). **ADULT MALE** Dark grey above with dark underparts, blackest on face and greyest on belly; orange-buff under tail. Wings brownish black with off-white panel. **ADULT FEMALE and IMMATURE** Mouse-grey, and tail-end colour less bright.

Song's quieter, strange ending sounds like crackling cellophane.

Summer visitor to SE England and France, eastwards to about 60°N and 40°E. Winters in W Europe, south to Mediterranean basin.

TOP: *JUVENILE*; ABOVE: *MALE*

FACT FILE

LENGTH 15cm WINGSPAN 23–26cm WEIGHT 13–20g FOOD Invertebrates, fruit EGGS 4–6; white, sometimes pale blue HABITAT Stony, craggy hillsides, regularly by farmsteads; recently, even towns. Often winters on rocky coasts VOICE Call: quiet *tsip*, often preceding *tuc tuc* alarm. Song: quick warble (*see also* 'Habits')

British Isles, especially in the W, and from France to Ukraine.

JAN | FEB | MAR | APR | MAY | JUN | JUL | AUG | SEP | OCT | NOV | DEC

STONECHAT

Saxicola torquata

ADULT MALE Distinctive black and white head. Back dark brown, white rump streaked brown. Wings dark brown with a variable-sized white panel. Breast orange, shading to white on belly. **ADULT FEMALE** Mottled brown head and upperparts, and less bright underparts. **JUVENILE** Like a heavily spotted female.

Mostly resident; N and E birds move S in winter. Nests on or near ground. Populations suffer in cold weather. Shuns intensive agriculture.

LEFT: *MALE;* ABOVE: *FEMALE*

FACT FILE

LENGTH 12.5cm WINGSPAN 18–21cm WEIGHT 11–19g FOOD Mostly insects EGGS 4–6, pale blue with some reddish markings HABITAT Gorse and other scrub, heath, sand-dunes, young plantations VOICE Call: *tchak*, like stones knocking together. Song: short, scratchy phrases

JAN | FEB | MAR | **APR** | **MAY** | **JUN** | **JUL** | **AUG** | **SEP** | **OCT** | NOV | DEC

WHINCHAT

Saxicola rubetra

Summer visitor to region (not Iceland), mostly further E and N than the Stonechat (opposite), into the far N of Scandinavia. Winters in sub-Saharan Africa.

ADULT MALE Distinctive dark brown head with a broad, long white supercilium. Dark brown upperparts, orange breast, white belly. Tail is black with white sides at base, noticeable in flight. **ADULT FEMALE and JUVENILE** Similarly patterned but duller, especially juvenile.

Nests on the ground. British numbers are down 26% since the 1990s, probably through loss of habitat in Africa. Song contains some phrases imitating those of other species.

ABOVE: *MALE;* RIGHT: *FEMALE*

FACT FILE

LENGTH 12.5cm WINGSPAN 21–24cm WEIGHT 13–26g FOOD Mainly invertebrates EGGS 4–7; blue, speckled red-brown HABITAT Open meadows and scrub; hilly, Bracken-covered slopes VOICE Call: harsh *tzee*. Song: long burst of phrases, some musical, some scratchy

JAN | FEB | MAR | APR | MAY | JUN | JUL | AUG | SEP | OCT | NOV | DEC

SONG THRUSH

Turdus philomelos

The most common and widespread thrush across the region except in the far N and Iceland. All migrate S or W in winter; British birds head S and are replaced by N and E birds.

Brown above, white below with buff tint to breast and flanks; underparts streaked (not spotted like Mistle Thrush, p. 114) with black-brown arrow-shaped spots. Orange-buff underwing, visible in flight.

Its nest of grass and moss is uniquely lined with mud and decayed wood. Breaks open snails on a stone.

FACT FILE

LENGTH 23cm
WINGSPAN 33–36cm
WEIGHT 65–100g
FOOD Invertebrates, fruit, snails
EGGS 4–6; bright blue, spotted black HABITAT Woodland, parks and gardens with plenty of shrubs
VOICE Call: *sipp*. Alarm: rapid *tchuk-tchuk-tchuk*. Song: loud, characteristic series of phrases, each often repeated 2–4 times

JAN | FEB | MAR | APR | MAY | JUN | JUL | AUG | SEP | OCT | NOV | DEC

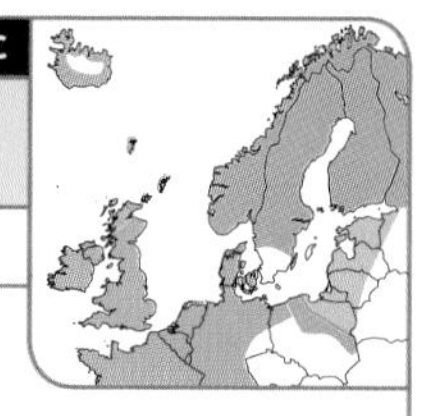

REDWING

Turdus iliacus

All upperparts dark, warm brown, darker on flight feathers. Breast yellowish buff on sides, with dark brown streaks. Underparts white, streaked with lighter brown. Flanks and underwing chestnut-red (*cf.* buff underwing of Song Thrush, p. 112).

Breeds in Iceland, across Scandinavia, around Baltic and into Siberia. British population numbers <100 pairs, compared with c. 1 million in Sweden. Winters in W and S Europe.

In winter, roosts in large flocks. Night migrants' calls keep the flock together and alert observers below.

FACT FILE

LENGTH 21cm WINGSPAN 33–34.5cm WEIGHT 50–80g FOOD Wide variety of invertebrates, also berries in autumn and winter EGGS 4–5; bluish, heavily speckled reddish brown HABITAT Open woods, thickets, scrub; winters on grassland, arable fields, open woodland VOICE Flight call: *see-ip*. Song: 4–6 fluty notes plus warbling

Across region except Iceland, but much less common in Norway. N and E populations move S and W in winter into milder parts of breeding range, forming small flocks.

JAN | FEB | MAR | APR | MAY | JUN | JUL | AUG | SEP | OCT | NOV | DEC

MISTLE THRUSH

Turdus viscivorus

ADULT Whitish underparts covered with large, wedge-shaped black spots; flanks and breast marked with buff. Upperparts and wings greyish brown with conspicuous greyish-white fringes. Tail grey-brown with diagnostic white tips to outer feathers. White underwing striking in flight. **JUVENILE** Spotted white on head, mantle and wing coverts.

Vigorously defends breeding territory. Song is far-carrying, even in wild weather.

FACT FILE

LENGTH 27cm WINGSPAN 42–47.5cm WEIGHT 100–140g FOOD Invertebrates, berries EGGS 4–5; pale blue, spotted reddish and purple HABITAT Orchards, woods, farmland, parks, gardens VOICE Call: harsh, distinctive rattle. Song: short, loud, fluty phrases

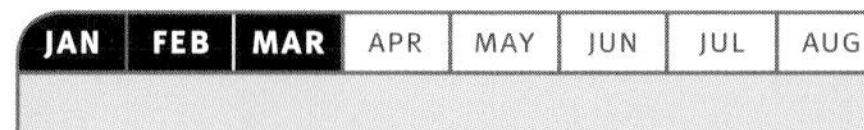

FIELDFARE

Turdus pilaris

Unmistakable upperparts: grey, chestnut, black. Throat and breast golden brown, streaked black. Rest of underparts white, flanks streaked black.

Gregarious. Many hundreds roost together in winter; most commonly breeds in colonies of <40–50 pairs. All feeding is on neutral ground. A noisy, aggressive bird in defence of its nest and winter food. The call readily identifies migrating flocks.

Breeds widely in Scandinavia, Finland and Germany, and occasionally elsewhere in region. N and E populations winter to W and S.

FACT FILE

LENGTH 25.5cm WINGSPAN 39–42cm WEIGHT 80–120g FOOD Many invertebrates, and fruits in autumn and winter EGGS 5–6, mostly light blue with brown markings HABITAT Open woodland, scrub, gardens, parks VOICE Call: *tchak tchak*. Song: weak warble with some wheezes and chuckles

JAN | FEB | MAR | APR | MAY | JUN | JUL | AUG | SEP | OCT | NOV | DEC

BLACKBIRD

Turdus merula

Breeds throughout, but only sparingly N of 65°N. N populations are migrants, moving to S and W of breeding range; others are resident.

ADULT MALE The only all-black European bird with a golden-yellow bill and long tail. Orange-yellow eye-ring. **FEMALE** All dark brown, underparts often with a rufous tone and thrush-like mottling on breast. **IMMATURE MALE** Plumage dull black and bill dark brownish horn. **JUVENILE** As female but more rufous, and more spotted below.

Hunts through leaf litter, digging with its bill.

ABOVE: *MALE;* TOP: *FEMALE*

FACT FILE

LENGTH 24–25cm WINGSPAN 34–38.5cm WEIGHT 80–140g FOOD Insects, earthworms, plus wild fruits in autumn and winter EGGS 4–5; light blue, thickly spotted with reddish brown HABITAT Most places where there are trees, but also on moors VOICE Call: *see*. Alarm: shrill chatter. Song: variety of fluty phrases

JAN FEB MAR **APR MAY JUN JUL AUG SEP OCT** NOV DEC

GARDEN WARBLER

Sylvia borin

Rounded head, stubby bill; brown upperparts (darker on wings); pale buff below, fading to white on belly and under-tail coverts.

Rare in gardens! Often sings from thick cover, so hard to see. Avoids competition with Blackcap (p. 120) by arriving later and feeding lower down. Song can be confused with Blackcap's but is pitched lower, with longer phrases.

Summer visitor. Widely distributed, except Iceland and N Scandinavia. W populations winter in W Africa, E populations in E and S Africa.

FACT FILE

LENGTH 14cm WINGSPAN 20–24.5cm WEIGHT 15–22g FOOD Mostly insects, plus some fruit and berries EGGS 4–5, whitish with variable markings HABITAT Broadleaf woodland with thick undergrowth, scrub and young conifers VOICE Call: loud *tac tac*. Song: rich, sustained warble

JAN	FEB	MAR	APR	**MAY**	**JUN**	**JUL**	**AUG**	**SEP**	**OCT**	NOV	DEC

LESSER WHITETHROAT

Sylvia curruca

Widespread from England and NE France, northwards to S Fennoscandia. Summer visitor, wintering mostly in NE Africa.

Beware confusion with the Whitethroat (opposite). Grey-brown above and white below. Dark face mask against slate-grey head. Primaries and tail browner and darker than upperparts. White outer tail feathers.

Skulking. Song's rattle may carry 200m and sound like the whole song; it is often preceded by a low warble, audible from only a few metres away.

FACT FILE

LENGTH 12.5–13.5cm WINGSPAN 16.5–20.5cm WEIGHT 10–19.5g
FOOD Mainly insectivorous, plus summer fruits EGGS 4–6; white, sparingly spotted olive HABITAT Woodland edges, thick hedges, shrubberies VOICE Call: hard *tack* or *churr*. Song: unlike any other warbler's – loud rattle and quiet warble

WHITETHROAT

Sylvia communis

ADULT MALE Grey cap to below eye, pure white chin and throat; breast pinkish buff, rest of underparts white; upperparts dull brown. Dark brown tail with white outer feathers. **ADULT FEMALE and JUVENILE** Same tail and wing patterns as male but duller plumage.

Suffered a population crash in 1968–69; numbers have since recovered slowly. Male builds several nests; female may complete one.

Summer visitor across Europe from 65°N in Fennoscandia southwards (but not in central region or Iceland). Winters mostly in tropical sub-Saharan Africa.

TOP: *JUVENILE*; ABOVE: *MALE*

FACT FILE

LENGTH 14cm WINGSPAN 18–23cm WEIGHT 11–21g FOOD Mostly insects; fruits in season EGGS 4–5, very pale blue with 'pepper and salt' spots HABITAT Open habitats with thickets and shrubs VOICE Call: scolding *charr* and sharp *tac*. Song: short, chattery warble, often in flight

JAN | FEB | MAR | APR | MAY | JUN | JUL | AUG | SEP | OCT | NOV | DEC

BLACKCAP

Sylvia atricapilla

Summer visitor to much of region (not Iceland), N to S Scandinavia. W European birds usually migrate to W Africa; the rest fly to E Africa.

Beware confusion with Marsh and Willow tits (p. 131). **ADULT MALE** Black crown, ash-grey nape and face. Upperparts ashy brown, tail and primaries darker. Chin, breast and flanks grey; belly and under-tail coverts white. **ADULT FEMALE and IMMATURE** Similar but cap bright red-brown and upperparts browner.

Forages and sings in treetops. Some winter in British Isles in urban parks and gardens.

RIGHT: *MALE*
BELOW: *FEMALE*

FACT FILE

LENGTH 13cm WINGSPAN 20–23cm WEIGHT 14–30g FOOD Insects and, especially in autumn, fruit and berries EGGS 4–6; very variable ground colour, with dark spots HABITAT Open woodland, copses with thick undergrowth, parks VOICE Call: loud, repeated *tac*. Song: loud, rich warble, rising in pitch

JAN | FEB | MAR | **APR** | **MAY** | **JUN** | **JUL** | **AUG** | **SEP** | **OCT** | NOV | DEC

SEDGE WARBLER

Acrocephalus schoenobaenus

Summer visitor, from high Arctic southwards. Absent from mountainous Scandinavia and Iceland. Winters in sub-Saharan Africa.

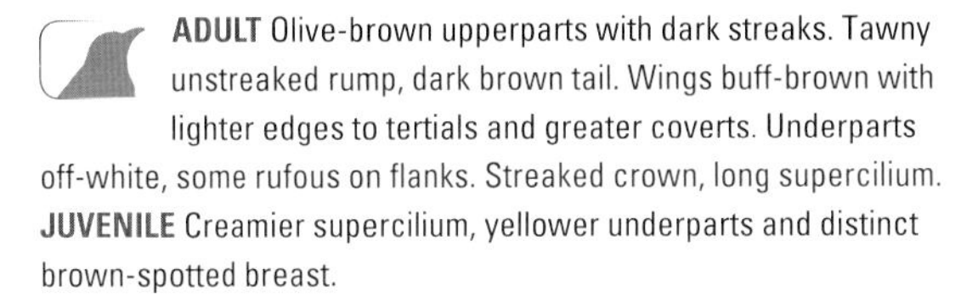

ADULT Olive-brown upperparts with dark streaks. Tawny unstreaked rump, dark brown tail. Wings buff-brown with lighter edges to tertials and greater coverts. Underparts off-white, some rufous on flanks. Streaked crown, long supercilium. **JUVENILE** Creamier supercilium, yellower underparts and distinct brown-spotted breast.

Birds gain weight rapidly on S migration at stopovers – some may double their weight and be able to fly non-stop from S Britain to S of Sahara.

FACT FILE

LENGTH 13cm WINGSPAN 17–21cm WEIGHT 8–21g FOOD Mostly insects EGGS 5–6; pale green, well speckled with olive HABITAT Low, lush vegetation near water VOICE Call: *tuc*. Alarm: *churr*. Song: loud mix of harsh and musical notes

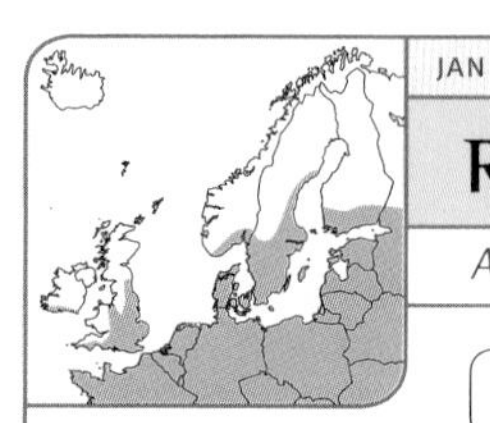

JAN FEB MAR **APR MAY JUN JUL AUG SEP** OCT NOV DEC

REED WARBLER

Acrocephalus scirpaceus

S Britain eastwards to around Baltic, where it is found mostly S of 60°N. Summer visitor, wintering in sub-Saharan Africa as far as Zambia.

Upperparts olive-brown, with a more rufous rump and upper-tail coverts, and darker brown wing-tips. Underparts white with buff under-tail coverts and sides of breast.

Its habitat is vulnerable, so managed conservation of this is necessary to preserve the species. Builds a unique nest woven around several stems. One of the main hosts of the Cuckoo (p. 77).

FACT FILE

LENGTH 13cm WINGSPAN 17–21cm WEIGHT 9–17g FOOD Insects EGGS 3–5; pale green, speckled and blotched olive HABITAT Reedbeds VOICE Call: *churr-churr*. Alarm: harsher. Song: low, guttural churring, with long phrases

JAN | FEB | MAR | APR | **MAY** | **JUN** | **JUL** | **AUG** | **SEP** | OCT | NOV | DEC

ICTERINE WARBLER

Hippolais icterina

Breeds from NE France eastwards, and to 60–65°N in S Fennoscandia and Russia. Summer visitor, wintering in Africa S of Equator. Rare annual visitor to British Isles.

ADULT Basically green above and yellow below, with a long bill, rather flat crown when relaxed and yellow face; long wings, reaching at least to upper-tail coverts. **LATE-SUMMER ADULT and JUVENILE** Browner above and whiter below.

Builds the most substantial nest of any warbler.

FACT FILE

LENGTH 13.5cm WINGSPAN 21–24cm WEIGHT 11–18g FOOD Insects, larvae, fruit EGGS 4–5; pale purplish pink, spotted with black HABITAT Sunny, wooded lowlands, cultivated land, gardens VOICE Call: *tec* and, in spring, a diagnostic *deeteroo*. Song: loud, remarkably long warble, lasting up to 40 seconds and carrying 500m

JAN FEB MAR **APR MAY JUN JUL AUG SEP OCT** NOV DEC

WILLOW WARBLER

Phylloscopus trochilus

Commonest leaf warbler in the region, occurring far into the Arctic in scrub beyond the tree-line, but not Iceland. Winters in tropical sub-Saharan Africa.

ADULT Olive-green above, yellowish white below; colours cleaner than in similar Chiffchaff (below). In summer, becomes browner above and whiter below from feather abrasion. **AUTUMN JUVENILE** Much yellower supercilium, throat and breast.

Active little bird, searching tree canopy for insects. Builds a well-concealed nest on ground.

FACT FILE

LENGTH 10.5–11.5cm WINGSPAN 17–22cm WEIGHT 7–13g FOOD Insects EGGS 4–8; white, finely spotted red HABITAT Woods, forests; mainly coppices and scrub VOICE Call: disyllabic, plaintive *hoo-eet*. Song: *c.* 3-second cascade of pure notes, dying away towards end

JAN FEB MAR APR MAY JUN JUL AUG SEP OCT NOV DEC

CHIFFCHAFF

Phylloscopus collybita

Second commonest leaf warbler in region, breeding throughout except Iceland and far N.

Like a less streamlined Willow Warbler (above), with a less yellow tint. Dull brownish olive above; dull, pale yellow below, shading to buff flanks. E forms are noticeably greyer above and whiter below. Pale yellow supercilium, pale eye-ring, contrasting dark eye.

MALE

Arrives as early as March. Most winter Mediterranean area and Africa S of Sahara; 500–1,000 winter in British Isles.

FACT FILE

LENGTH 10–11cm WINGSPAN 15–21cm WEIGHT 7–10g FOOD Insects EGGS 4–7; white, sparingly spotted purplish HABITAT Open woodland with a good shrub layer, copses, hedgerows VOICE Call: monosyllabic *hweet, hweet*. Song: diagnostic variations on *chiff-chaff-chiff-chaff*

JAN | FEB | MAR | APR | MAY | JUN | JUL | AUG | SEP | OCT | NOV | DEC

GOLDCREST

Regulus regulus

MALE

ADULT Tiny, with dull greenish upperparts, 2 white wing bars and pale olive-green underparts, darker on flanks. Crown on both sexes yellow, lined with black; displaying male raises crest to reveal orange centre. **JUVENILE** Lacks crown stripe.

Suffers severely in bad winters, but populations can recover well. Tiny size, restlessness and high-pitched voice often make it difficult to spot. Some people cannot hear its calls.

Widespread resident, except Iceland; migrates S and W in winter, from only the harshest weather.

FACT FILE

LENGTH 9cm WINGSPAN 13.5–15.5cm WEIGHT 4.5–7.5g FOOD Insects EGGS 4–7, white with brownish speckles HABITAT Breeds in mature conifers; will inhabit deciduous woods, gardens and parks if they contain suitable conifers VOICE Call: *zit, zit*. Song: sweet-sounding repeated pattern

JAN | FEB | MAR | APR | MAY | JUN | JUL | AUG | SEP | OCT | NOV | DEC

FIRECREST

Regulus ignicapillus

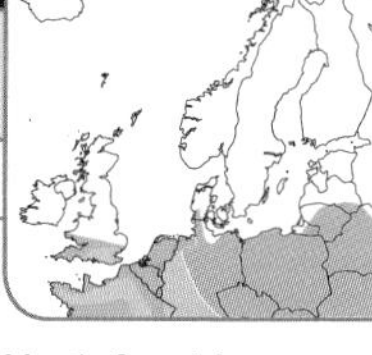

Tiny. Golden crown outlined with black, and broad white supercilium underlined by black eye-stripe; displaying male raises crown to reveal orange 'fire crest'; female's crown is yellow. Upperparts bright olive-green, with 2 white wing bars; tail is darker and browner; all underparts are white.

Spends more time in broadleaved trees and shrubs than Goldcrest (above), and prefers larger prey than Goldcrest.

Mostly found from France to Poland. N and E populations winter in Mediterranean basin and far W. Breeding range has expanded W. First bred in England in 1962.

FACT FILE

LENGTH 9cm WINGSPAN 13–16cm WEIGHT 4–6.5g FOOD Especially arthropods EGGS 7–12, pinkish buff with tiny red dots HABITAT Trees (mostly broadleaved) and shrubs VOICE Call: very high-pitched *zit zit*, lower than Goldcrest's (p. 125). Song: rapid string of calls

JAN | FEB | MAR | **APR** | **MAY** | **JUN** | **JUL** | **AUG** | **SEP** | **OCT** | NOV | DEC

SPOTTED FLYCATCHER

Muscicapa striata

Summer visitor throughout except Iceland. Winters in Africa, mostly S of Equator.

ADULT Upperparts, wings and tail plain grey-brown. Forehead and crown streaked black, outlined in white. Underparts white, washed with brown on side of breast and flanks. **JUVENILE** Upperparts buffer than adult's; head, back and wings have pale, round buff-white spots. Underparts not streaked but spotted dark brown. The juvenile gives the bird its name.

Often the last visitor to arrive. Population has declined by 75% since the 1960s. Obtains most of its food in flight, sallying forth from a perch.

FACT FILE

LENGTH 14.5cm WINGSPAN 23–25.5cm WEIGHT 10–20g FOOD Aerial insects EGGS 4–6; off-white, blotched reddish and purplish HABITAT Woodland edges, parks, orchards, gardens VOICE Call: *tzee-zuk-zuk*. Song: quiet and short, like a squeaky wheelbarrow

JAN | FEB | MAR | **APR** | **MAY** | **JUN** | **JUL** | **AUG** | **SEP** | **OCT** | NOV | DEC

PIED FLYCATCHER

Ficedula hypoleuca

BREEDING MALE Unmarked white below; black above, relieved by white forehead; white-edged tertials, which meet white wing bar; white on outer tail feathers. **NON-BREEDING MALE, ADULT FEMALE and JUVENILE** Upperparts brown or grey, tail and rump darker.

Summer visitor to W Britain, and widely from Germany eastwards into N Scandinavia. Winters in W African forests.

Readily uses nestboxes, so has been studied in detail. Many males successively have 2 or 3 mates, usually feeding just the 1st brood. Widespread in open woodland.

TOP: *FEMALE;* ABOVE: *MALE*

FACT FILE

LENGTH 13cm WINGSPAN 21.5–24cm WEIGHT 10–16g FOOD Flying and non-flying arthropods EGGS 6–7, pale blue HABITAT Deciduous and mixed open woodland VOICE Call: loud *whit* or *wee-tic*. Song: rapid sequence of high and low notes

JAN | FEB | MAR | APR | MAY | JUN | JUL | AUG | SEP | OCT | NOV | DEC

GREAT TIT

Parus major

The world's most widespread tit, from British Isles to Far East. Sedentary in Britain. N and E European birds migrate S or W. Breeds in territories; winters in flocks.

Black and white head; yellow underparts with black centre stripe, female's narrower than male's; greenish back; blue-grey wings and tail, with 1 white wing bar and white outer tail.

Regular at birdfeeders and nestboxes. Much studied.

FACT FILE

LENGTH 14cm WINGSPAN 22.5–25.5cm WEIGHT 14–22g FOOD Insects, spiders and their larvae; also seeds and fruit in winter EGGS 3–18; white, variably spotted reddish brown HABITAT Woodland, parkland, orchards, hedgerows, town gardens VOICE Call: sharp *chink*. Male usually has several distinct songs based on typical *tea-cher*

JAN	FEB	MAR	APR	MAY	JUN	JUL	AUG	SEP	OCT	NOV	DEC

BLUE TIT

Cyanistes caeruleus

Widespread resident to c. 65°N (not Iceland). Most are likely to move <10km. Populations erupt in years of high numbers.

Region's only tit with a blue crown, outlined in white. **ADULT** Dark line through eye; cheeks white, outlined in black from chin. Upperparts yellowish green. Underparts sulphur-yellow. Wings and tail dark blue. **JUVENILE** Similar but face washed yellow.

Abundant. Wanders outside breeding season. Feeds high in trees, and is a regular visitor to bird tables in winter; agile in order to reach food. Readily uses nestboxes.

FACT FILE

LENGTH 11.5cm WINGSPAN 17.5–20cm WEIGHT 9–12g FOOD Insects, defoliating caterpillars; fruits and seeds in season EGGS 10–12, white with reddish spots HABITAT Trees, even inner-city parks and gardens. Avoids conifers VOICE Call: tsee-tsee. Alarm: *chirr-r-r*. Song: *tsee-tsee-tsee-tsuhuhuhu* tremolo

Widespread resident to about 65°N (not Iceland), but generally less common than other tits.

JAN | FEB | MAR | APR | MAY | JUN | JUL | AUG | SEP | OCT | NOV | DEC

COAL TIT

Periparus ater

Glossy black cap; large white patch on nape. White cheeks. Chin, throat and upper breast black. Underparts buff, paler towards centre. Upperparts, wings and tail olive-grey. Two white wing bars.

The tit of coniferous woods, but also occurs in other woodland. Many seeds are stored in a crevice for retrieval later. More agile than other tits when foraging. Nests in hole in a tree, wall or rock.

FACT FILE

LENGTH 11.5cm WINGSPAN 17–21cm WEIGHT 9–11g FOOD Insects and spiders, plus seeds in autumn and winter EGGS 8–9; white, finely spotted red-brown HABITAT Anywhere there are firs – even urban cemeteries, parks and gardens VOICE Call: piping *tsee*. Song: loud, clear *teechu, teechu, teechu*

JAN | FEB | MAR | APR | MAY | JUN | JUL | AUG | SEP | OCT | NOV | DEC

MARSH TIT

Poecile palustris

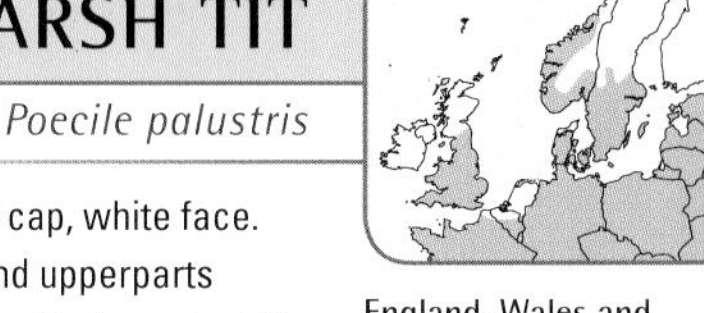

Glossy black cap, white face. Wings, tail and upperparts greyish brown. Underparts dull white, with pale buff tinge on flanks and under-tail coverts. Separable from Willow Tit (below) with difficulty – note glossy black cap, small black bib with well-defined edges, paler underparts, and quite distinct calls and song.

England, Wales and SE Scotland, thence eastwards to Baltic and S Scandinavia.

Resident, spending all year in the same territory. Does not often feed at bird tables; nests in natural holes.

FACT FILE

LENGTH 11.5cm WINGSPAN 18–19.5cm WEIGHT 9–13g FOOD Invertebrates, plus seeds and fruits in winter EGGS 7–10, white with reddish-brown spots at larger end HABITAT Deciduous woodland, not marshes! VOICE Call: *pitchoo* and nasal *ter-char-char-char*. Song: repetition of 1 note

JAN | FEB | MAR | APR | MAY | JUN | JUL | AUG | SEP | OCT | NOV | DEC

WILLOW TIT

Poecile montanus

Dull black cap extends to mantle. Bib has poorly defined borders. White cheeks, whitish underparts, greyish-brown upperparts, whitish wing patch. Always observe plumage and call notes to be sure of identification (*cf.* Marsh Tit above).

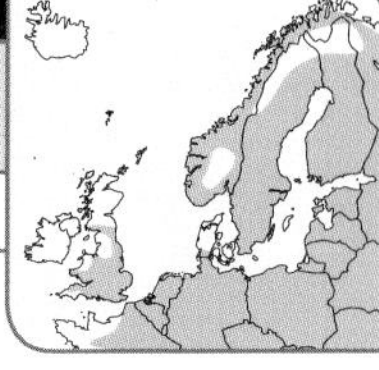

S Scotland, England, Wales and E France, eastwards to Scandinavia, Finland and Russia.

Resident. In Britain, found regularly only in damp lowland. Often the commonest tit in N Europe. Noteworthy for male and female excavating their own nest-hole, low in a rotten stump.

FACT FILE

LENGTH 11.5cm WINGSPAN 17–20.5cm WEIGHT 9.5–13g FOOD Invertebrates in breeding season; otherwise, seeds and berries EGGS 6–9; white, spotted red-brown HABITAT Mixed woodland, coniferous forest, trees in damp lowland VOICE Call: *eez-eez-eez* and characteristic nasal *tchay, tchay*

Widespread to c. 65°N except in Britain, where it is confined to N Scotland. Not Iceland.

JAN | FEB | MAR | APR | MAY | JUN | JUL | AUG | SEP | OCT | NOV | DEC

CRESTED TIT

Lophophanes cristatus

Crest is backward-pointing; black, tipped white. **ADULT** White face with curving black line from eye; black line down side of neck to join black bib. Upperparts buff-brown, wings and tail grey-brown. **JUVENILE** Has shorter crest.

Very sedentary; less common than other tits. Its range is further restricted by its need for rotten wood in which to excavate its nest-hole. Stores food in autumn for use in winter.

FACT FILE

LENGTH 11.5cm WINGSPAN 17–20cm WEIGHT 10–12.5g FOOD Insects, spiders; seeds in season EGGS 6–7; white, spotted red HABITAT Pine forest in the N; mixed or deciduous woods elsewhere VOICE Limited vocal repertoire: low-pitched, purring trill; song using repeated calls

JAN | FEB | MAR | APR | MAY | JUN | JUL | AUG | SEP | OCT | NOV | DEC

LONG-TAILED TIT

Aegithalos caudatus

Widespread resident across Europe, except Iceland and much of N Fennoscandia.

Head and underparts whitish, washed pink. Upperparts, wings and tail dull black; scapulars and rump pink; 9cm-long tail has white tips and edges. Subspecies in N (*A. c. caudatus*) and E (*A. c. europaeus*) have a pure white head and white-edged wings; W subspecies (*A. c. rosaceus*) has black stripe from bill to mantle.

Breeding pairs are territorial, taking *c.* 3 weeks to build a domed nest of moss covered with lichen and lined with *c.* 1,000 feathers.

ABOVE: *ADULT BRITISH RACE* (A. c. rosaceus)
TOP: *NORTHERN RACE* (A. c. caudatus)

FACT FILE

LENGTH 14cm WINGSPAN 14–19cm WEIGHT 7–9g FOOD Bugs, plus insect eggs and larvae EGGS 8–12; white, with some minute reddish spots HABITAT Deciduous woodland, thick scrub VOICE Call: *tsirrrup*. Song: rapid repetition of calls

JAN | FEB | MAR | APR | MAY | JUN | JUL | AUG | SEP | OCT | NOV | DEC

NUTHATCH

Sitta europaea

S Norway, S Sweden and N Russia, westwards to Atlantic coast and S Britain. Not Ireland.

Upperparts blue-grey, cheeks and throat white; rest of underparts orange-buff, but Fennoscandia birds paler. Broad black eye-stripe. Outer tail feathers black with white sub-terminal spots.

Sedentary. Regularly visits bird tables. Only bird in region that can move head-first down a tree trunk. Female only builds the nest in a tree-hole or box, plastering entrance with mud to reduce its size.

FACT FILE

LENGTH 14cm WINGSPAN 22.5–27cm WEIGHT 22–25g FOOD Invertebrates; seeds in winter EGGS 6–8; white, sparsely marked reddish HABITAT Broadleaf and mixed woodland, gardens with mature trees VOICE Call: loud *chwit-chwit*. Song: rapid *chu-chu-chu* and slow *pee, pee, pee*

JAN | FEB | MAR | APR | MAY | JUN | JUL | AUG | SEP | OCT | NOV | DEC

TREECREEPER

Certhia familiaris

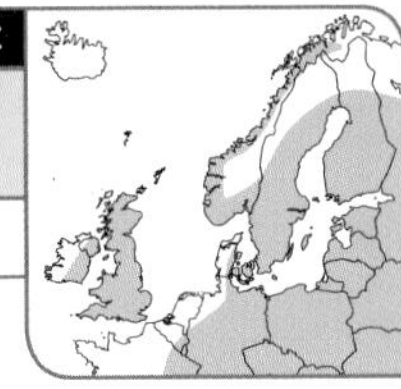

Brown above and white below, relieved by rufous rump, white supercilium, mottled and streaked back, 2 pale wing bars, distinctive buff band across wing, and white-spotted tertials. Tail long and brown. Bill quite long.

Hard to observe because of its cryptic colours and high-pitched voice. Searches for food up a tree, then flies to low down on another.

Widely distributed from E France eastwards and throughout British Isles. Resident, but N populations move to winter within breeding range. Numbers are hard hit by prolonged frosts.

FACT FILE

LENGTH 12.5cm WINGSPAN 17.5–21cm WEIGHT 8.5–10.5g FOOD Insects, spiders EGGS 5–6, white with reddish spots HABITAT Coniferous forest in Continental Europe; deciduous forest in British Isles VOICE Call: thin *tsiew*. Song: cadence lasting 2.5–3 seconds

JAN | FEB | MAR | APR | MAY | JUN | JUL | AUG | SEP | OCT | NOV | DEC

MAGPIE

Pica pica

Sedentary throughout Europe. Adults may spend all their lives in the same territory; 1st-year birds disperse.

Unmistakable: tail comprises over half total length. **ADULT** Scapulars, outer half of wings and flanks white; the rest black with iridescence. Distinctive pied pattern in flight. **JUVENILE** Shorter tail, duller plumage.

As numbers have increased, it has moved into suburban and urban habitats in several countries. Feeds on ground. Recent research has shown the breeding success of small birds has not been harmed by increasing numbers.

FACT FILE

LENGTH 44–46cm WINGSPAN 52–60cm WEIGHT 160–240g FOOD Invertebrates in summer, vertebrates and seeds in winter, carrion and scraps EGGS 5–7; variable ground colour, heavily speckled brown HABITAT Lightly wooded lowlands VOICE Loud, staccato *chacker chacker chacker chacker*

JAY

Garrulus glandarius

Our most colourful crow. Pinkish-brown body with a white rump and under-tail coverts; black tail; white forehead and crown, streaked black. Wings black with a short white bar, and shiny blue bars on shoulders.

W and S birds are sedentary; others are eruptive migrants when the acorn crop fails.

From within Arctic Circle to Mediterranean, and from Atlantic coast to Japan. Has increased in Britain, even moving into urban areas

FACT FILE

LENGTH 34–35cm WINGSPAN 52–58cm WEIGHT 145–190g FOOD Invertebrates, fruits, seeds; acorns form staple winter diet EGGS 5–7; pale greenish, densely spotted olive-brown HABITAT Fairly dense woodland, usually broadleaf VOICE Call: loud, raucous *skaaak skaaak*

JAN	FEB	MAR	APR	MAY	JUN	JUL	AUG	SEP	OCT	NOV	DEC

NUTCRACKER

Nucifraga caryocatactes

S Fennoscandia to about 60°N, E side of Baltic, and N and E into Russia to above 65°N; another population is found in suitable habitat SE from Belgium and Germany.

Mostly chocolate-brown head and body, streaked with white spots. Plain, dark brown cap. Wings blackish brown with small white spots on coverts. Tail white below and in corners.

Resident. W birds in winter eat Hazel nuts, normally caching *c.* 100,000 seeds. If N crop fails, hundreds or thousands erupt westwards, even reaching British Isles. In many areas birds start breeding when thick snow is still on ground. Quite noisy outside breeding season.

FACT FILE

LENGTH 22–33cm WINGSPAN 52–58cm WEIGHT 100–190g
FOOD Mainly conifer seeds EGGS 3–4; greenish blue, finely spotted olive HABITAT Coniferous forest VOICE High-pitched *kraak*

JAN | FEB | MAR | APR | MAY | JUN | JUL | AUG | SEP | OCT | NOV | DEC

JACKDAW

Corvus monedula

Small black crow with a short bill. Grey nape (palest among E birds), with clear-cut edges between crown and mantle.

Needs some open ground on which to feed. Lifelong pairing. Nests in small colonies in tree-holes, crevices in cliffs, nestboxes, and holes in church towers and other buildings. In non-breeding season, roosts in large numbers.

Mostly resident across region to 60–65°N, but rare in Norway and absent from Iceland. E birds migrate W to winter in the breeding range.

FACT FILE

LENGTH 33–34cm WINGSPAN 67–74cm WEIGHT 185–280g FOOD Invertebrates, seeds, fruit, carrion EGGS 4–6, light blue or greenish with variable darker speckles HABITAT Old woodland, parks, coastal cliffs, quarries, urban areas VOICE Call: sharp *chack*

JAN | FEB | MAR | APR | MAY | JUN | JUL | AUG | SEP | OCT | NOV | DEC

ROOK

Corvus frugilegus

British Isles and France eastwards into Russia; scattered populations in S Fennoscandia. Resident in British Isles; other N and E birds migrate S and W in winter.

ADULT All black. Sharp, pointed black bill with whitish grey at base, and steep forehead. **JUVENILE** Black at base of bill.

Tall trees are needed for colonial nesting in 'rookeries'; these may contain hundreds of nests. Also gregarious outside the breeding season for feeding, roosting (roosts may contain thousands of birds) and on migration.

FACT FILE

LENGTH 44–46cm WINGSPAN 81–99cm WEIGHT 370–570g FOOD Earthworms, beetles, cereal seeds EGGS 2–6; light blue or green, spotted with olive HABITAT Agricultural country with some tall trees VOICE Call: *kaah*, less harsh than that of Carrion or Hooded crows (p. 141)

JAN | FEB | MAR | APR | MAY | JUN | JUL | AUG | SEP | OCT | NOV | DEC

CARRION CROW/HOODED CROW

Corvus corone/Corvus cornix

Unlike Rook (p. 140), both lack white at base of bill, and have a less steep forehead and more square-ended tail. **CARRION CROW** All black. **HOODED CROW** Contrasting grey back and underparts.

Breed in a wide variety of habitats. Even search for food on estuary mud, rubbish tips and in cities. Walk with a steady gait. Nest singly, usually in a tree. Both considered to be vermin.

CARRION CROW resident in England, Wales and France, eastwards to Germany and Czech Republic (green on map). HOODED CROW resident in Ireland, Isle of Man, NW Scotland and its islands, Faeroes, Denmark and Fennoscandia (blue and green hatching on map). Hybrids with great variations in colour patterns occur at borders of distribution.

TOP: *HOODED CROW*
LEFT: *CARRION CROW*

FACT FILE

LENGTH Both 45–47cm WINGSPAN 93–104cm WEIGHT 400–640g FOOD Almost anything! EGGS 3–6; blue or green, spotted olive and brown HABITAT Coast, arable land, grassland, parkland, heaths, moors VOICE Call: hoarse *kraah*

Iceland, Faeroes, W British Isles, W France and Denmark, eastwards and northwards to Arctic.

JAN | FEB | MAR | APR | MAY | JUN | JUL | AUG | SEP | OCT | NOV | DEC

RAVEN

Corvus corax

The largest crow, one-third larger than Carrion/Hooded crows (p. 141) or Rook (p. 140). All black with a wedge-shaped tail, massive bill, flat head and long wings. In flight, heavy head and bill help to give it a cruciform silhouette.

Needs an undisturbed nest site on a sea cliff, quarry or tree. Powerful flight; soars freely and performs aerobatics (diving, flipping on its back), mainly in breeding season. Ranges widely for food.

FACT FILE

LENGTH 64cm WINGSPAN 120–150cm WEIGHT 900–1,400g FOOD Varies, but especially carrion EGGS Light blue or blue-green, with variable dark markings HABITAT Sea-level to high mountains, usually avoiding dense forest and farmland VOICE Repeated, deep *pruk* or *kronk*

JAN | FEB | MAR | APR | MAY | JUN | JUL | AUG | SEP | OCT | NOV | DEC

STARLING

Sturnus vulgaris

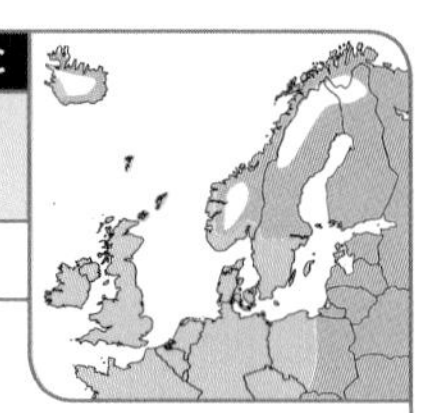

Breeds throughout region except high Arctic. N and E populations winter in the milder W.

Dumpy, short-tailed; walks with a waddle, not the Blackbird's hop (p. 116); has a delta-winged shape in flight. **SUMMER ADULT** Black with green and purple sheen. **WINTER ADULT** Covered with whitish spots. **JUVENILE** Dull, dirty brown with whitish throat; moults to adult winter plumage, the head the last part to change.

Digs for invertebrates with pointed bill. Nests in holes in buildings and trees. Winter roosts preceded by spectacular flights, sometimes of thousands of birds.

ABOVE: *WINTER;* RIGHT: *SUMMER*

FACT FILE

LENGTH 21.5cm WINGSPAN 37–42cm WEIGHT 70–95g FOOD Mostly invertebrates; also caterpillars, seeds, fruit EGGS 5–7, pale blue HABITAT Farmland and suburbs, especially fields and lawns with short grass VOICE Usual call: grating *cherrr*. Song: series of whistles, squeaks, warbles and mimicry

JAN | FEB | MAR | APR | MAY | JUN | JUL | AUG | SEP | OCT | NOV | DEC

HOUSE SPARROW

Passer domesticus

Found throughout – even a few in Iceland and Arctic Fennoscandia. Sedentary.

MALE Upperparts and wings chestnut, streaked black; crown and nape grey; black bib with broken bottom edge; dirty white cheeks; grey rump, dark brown tail; 1 white wing bar; rest of underparts dull grey. **FEMALE** Brown above, all underparts dull grey; olive-brown crown; pale buff supercilium, especially noticeable behind eye.

Ground feeder. Nests in small colonies, in holes in trees, buildings and nestboxes, or a domed nest in a shrub or tree. Sociable.

ABOVE: *MALE;* TOP: *FEMALE*

FACT FILE

LENGTH 14–15cm WINGSPAN 21–25.5cm WEIGHT 25–38g FOOD Wild seeds, cereal stubble, insects and their larvae for their young, food at bird tables EGGS 3–5; white, marked grey and brown HABITAT Close to farmland, towns and cities VOICE Noisy chirruping

JAN | FEB | MAR | APR | MAY | JUN | JUL | AUG | SEP | OCT | NOV | DEC

TREE SPARROW

Passer montanus

Fennoscandia to just above 60°N, eastwards to Russia and westwards to British Isles (mostly in E and central England). Mainly sedentary, but N populations sometimes erupt.

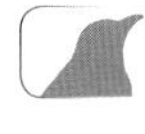

Similar to House Sparrow (p. 144), but note red-brown cap, small black bib with sharp bottom edge, white cheeks and incomplete collar, black patch below and behind eye, and double white wing bars. Sexes alike.

Nests in holes in trees, buildings and nestboxes. Winters on stubble, in farmyards. Less associated with Man in the W.

FACT FILE

LENGTH 14cm WINGSPAN 20–22cm WEIGHT 19–29g FOOD Invertebrates, seeds EGGS 4–6; whitish, heavily marked dark brown HABITAT Open country with mature trees, orchards, old hedgerows, pollarded willows by slow-flowing rivers, locally in gardens VOICE Call: sharp, repeated *teck*, higher-pitched than House Sparrow's

JAN | FEB | MAR | APR | MAY | JUN | JUL | AUG | SEP | OCT | NOV | DEC

CHAFFINCH

Fringilla coelebs

Widespread in British Isles and eastwards into Fennoscandia and Russia; not Iceland. N and E birds migrate, wintering mostly within breeding range.

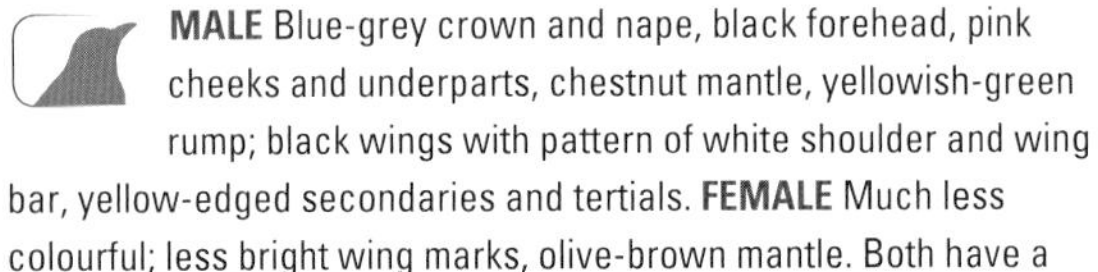

MALE Blue-grey crown and nape, black forehead, pink cheeks and underparts, chestnut mantle, yellowish-green rump; black wings with pattern of white shoulder and wing bar, yellow-edged secondaries and tertials. **FEMALE** Much less colourful; less bright wing marks, olive-brown mantle. Both have a black tail with white outer feathers.

Common. Territorial breeder; builds a beautiful cup-shaped moss nest, lined with hair, in a bush. Gregarious in winter.

male: multicoloured

olive-brown mantle

female: buffish grey

ABOVE: *MALE*
RIGHT: *FEMALE*

FACT FILE

LENGTH 14.5cm WINGSPAN 24.5–28.5cm WEIGHT 20–30g FOOD Seeds; caterpillars for young EGGS 4–5, bluish with reddish and purplish blotches HABITAT Deciduous and coniferous woodland, parks, gardens VOICE Call: metallic *chink*; male's spring call a monotonous, repeated *wheet*. Song: tuneful, ending in a flourish

JAN FEB MAR APR MAY JUN JUL AUG SEP OCT NOV DEC

BRAMBLING

Fringilla montifringilla

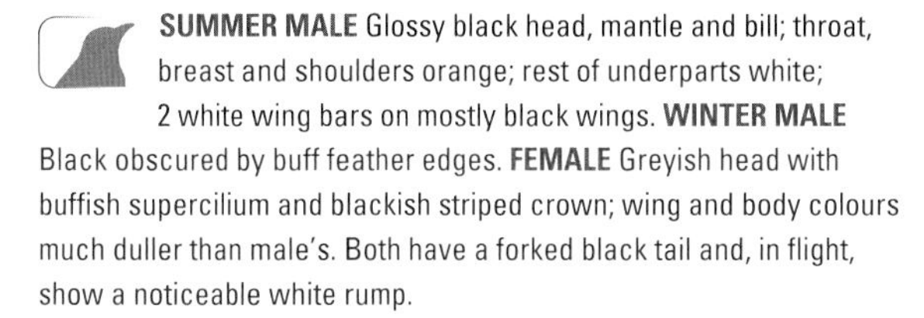

SUMMER MALE Glossy black head, mantle and bill; throat, breast and shoulders orange; rest of underparts white; 2 white wing bars on mostly black wings. **WINTER MALE** Black obscured by buff feather edges. **FEMALE** Greyish head with buffish supercilium and blackish striped crown; wing and body colours much duller than male's. Both have a forked black tail and, in flight, show a noticeable white rump.

Winters especially in Beech woods, feeding on fallen seed, and on stubble, often with other finches.

Summer visitor to Fennoscandia from 60°N northwards, and E across Russia. Winters S to Mediterranean and W to British Isles.

TOP: *FEMALE;* ABOVE: *MALE*

FACT FILE

LENGTH 14cm WINGSPAN 25–26cm WEIGHT 20–30g FOOD Summer, invertebrates; winter, seeds EGGS 5–7, variable with rusty spots HABITAT Open birch and mixed forest VOICE Call: tinny *tsweep*. Song: drawn out and wheezy

JAN	FEB	MAR	APR	MAY	JUN	JUL	AUG	SEP	OCT	NOV	DEC

LINNET

Carduelis cannabina

British Isles, across to Fennoscandia, around Baltic and into Russia. In winter, N and E birds migrate S and W; W birds migrate S as far as Iberia.

SUMMER MALE Smart grey, chestnut and crimson bird; white edges to blackish primaries and tail show well in flight. **ADULT FEMALE** Brown above, streaked darker; buff below with brown streaks; white belly; tail and wings have diagnostic white like summer male. **WINTER MALE and JUVENILE** Similar to adult female.

Sings from an open perch. Large winter flocks feed on stubble and fallow ground.

LEFT: *MALE:* ABOVE: *FEMALE*

FACT FILE

LENGTH 13.5cm WINGSPAN 21–25.5cm WEIGHT 15–20g FOOD Seeds EGGS 4–6, pale blue with reddish spots HABITAT Heaths, commons, coastal scrub, young plantations VOICE Flight call: *chichichit*. Sings a pleasant twitter

JAN | FEB | MAR | APR | MAY | JUN | JUL | AUG | SEP | OCT | NOV | DEC

LESSER REDPOLL/COMMON REDPOLL

Carduelis cabaret/Carduelis flammea

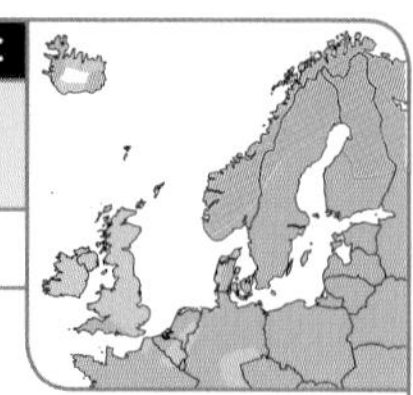

LESSER Overall buff tone; brown above, white below, and heavily streaked on wings, breast and flanks; buff wing bars. Small, pointed bill; tiny black bib; pinkish-red forehead. Summer males have a red breast. **COMMON** Very similar but paler and grey, with whitish wing bars.

Both are gregarious outside breeding season.

LESSER breeds in British Isles and along North Sea coast; COMMON breeds in Iceland and Fennoscandia eastwards. LESSER winters mostly in British Isles; COMMON migrates S and SE, some reaching France and Britain each year.

LEFT: *LESSER REDPOLL MALE*
BELOW LEFT: *COMMON REDPOLL MALE*

FACT FILE

LENGTH Lesser 11.5cm/Common 12.5cm WINGSPAN <22.5cm/>21cm WEIGHT 9–12g/10–15g FOOD Mainly seeds EGGS 4–6; pale blue, spotted and streaked HABITAT Willow, birch, Alder and Juniper forest; conifers, especially in British Isles VOICE Distinctive call: twittering *chuch-uch-uch-uch*

JAN | FEB | MAR | APR | MAY | JUN | JUL | AUG | SEP | OCT | NOV | DEC

GOLDFINCH

Carduelis carduelis

Not Iceland or N Scotland, but occurs in rest of British Isles eastwards to S Fennoscandia, Baltic states and Russia. Winters within the breeding area, as far S as Mediterranean.

Unmistakable. **ADULT** Red–white–black-striped head, the black reaching onto crown and the white joining under red chin; upperparts, breast and flanks sandy brown; rump white; white-tipped black tail slightly forked; wings black with a broad golden-yellow band right across and white-tipped flight feathers. Noticeably pointed bill. **JUVENILE** Greyish-buff head and body; wings and tail as adult.

Gregarious outside breeding season.

ABOVE: *ADULT;* TOP: *JUVENILE*

FACT FILE

LENGTH 12cm WINGSPAN 21–25.5cm WEIGHT 13–20g FOOD Weed seeds, especially thistles EGGS 5–6, bluish white with some reddish spots HABITAT Orchards, gardens, fringes of woods and commons VOICE Call: trisyllabic *tswitt-witt-witt*, like tiny bells. Song: also tinkling

JAN | FEB | MAR | APR | MAY | JUN | JUL | AUG | SEP | OCT | NOV | DEC

GREENFINCH

Carduelis chloris

Not Iceland; otherwise across region to Fennoscandia and Russia (mostly S of 65°N). Most are short-distance migrants, although N birds cover 1,000km or more.

ADULT MALE Olive-green above and yellowish green below, with grey patch on base of secondaries and yellow base to primaries; blackish tail with bright yellow base. **ADULT FEMALE** Duller and browner above, and paler and greyer below. **JUVENILE** Like female, and heavily streaked above and below.

Feeds on ground; can even crack open a Hornbeam nut. Circular song flight with fluttering wings.

TOP: *FEMALE*; ABOVE: *MALE*

FACT FILE

LENGTH 15cm WINGSPAN 24.5–27.5cm WEIGHT 25–35g FOOD Variety of seeds EGGS 4–6; bluish white, speckled and spotted HABITAT Parks, gardens, cemeteries, borders of woods VOICE Male's call: drawn-out, nasal *tswee*. Flight call: *teu teu*

SISKIN

Carduelis spinus

Breeds mostly in N and W British Isles, Scandinavia and around Baltic, eastwards into Russia. Winters extensively in W Europe.

ADULT MALE Yellow face (with greenish ear coverts), breast, rump and outer tail. Crown and chin black. Wing an intricate pattern of black, bright yellow and green, with 1 particularly broad yellow bar. **ADULT FEMALE** Patterned as male but duller, and without black cap. **JUVENILE** Like a dull female but more heavily streaked.

Searches acrobatically in trees for food. Winter visitor to bird tables.

LEFT: *JUVENILE*; RIGHT: *MALE*

FACT FILE

LENGTH 12cm WINGSPAN 20–23cm WEIGHT 11–16g FOOD Tree seeds, especially birch, Alder and spruce EGGS 3–5, bluish with dark spots and blotches HABITAT Coniferous and mixed forest, especially in river valleys VOICE Call, often in flight: clear *tsüü*. Song: non-stop twitter, ending in a wheeze

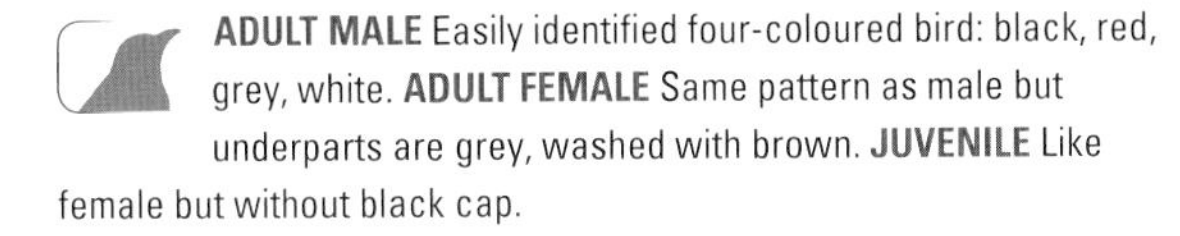

BULLFINCH

Pyrrhula pyrrhula

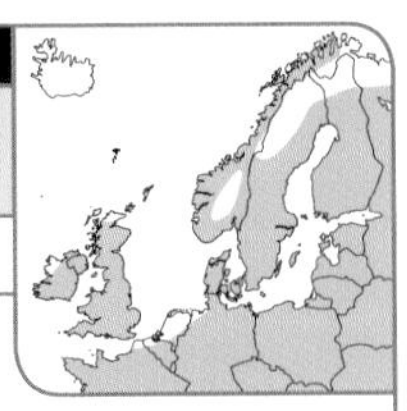

Throughout region except very far N and Iceland. Mostly resident; some of the most N birds migrate into S Scandinavia and central Europe.

ADULT MALE Easily identified four-coloured bird: black, red, grey, white. **ADULT FEMALE** Same pattern as male but underparts are grey, washed with brown. **JUVENILE** Like female but without black cap.

Quiet, unobtrusive species, whose song is mostly so quiet many birdwatchers have never heard it!

LEFT: *MALE;* RIGHT: *FEMALE*

FACT FILE

LENGTH 14.5–16.5cm WINGSPAN 22–29cm WEIGHT 25–35g FOOD Buds, berries, seeds EGGS 4–5; pale greenish blue with some purple-brown spots HABITAT Mixed woods, parks, gardens and, especially in Fennoscandia, coniferous forest VOICE Call note: subdued *phew*, falling in pitch. Slow, low-pitched song

JAN | FEB | MAR | APR | MAY | JUN | JUL | AUG | SEP | OCT | NOV | DEC

COMMON CROSSBILL

Loxia curvirostra

Mostly Fennoscandia and Baltic states eastwards; scattered populations W to British Isles. Resident, but eruptive some years. Red area on map shows distribution of Scottish Crossbill.

ADULT MALE Brick-red head and body, brown wings. Bill is longer than deep, with unique crossed mandibles, the upper curved more than the lower. **ADULT FEMALE** Grey-green with diffuse streaked upperparts, and yellower rump. **IMMATURE MALE** Often not completely red. **JUVENILE** Like a more heavily streaked and browner female. Very similar **Scottish Crossbill** (*L. scotica*), resident in pine forests in NE Scotland, has a deeper bill.

Cone's scales are opened with mandible tips and seed is extracted with the tongue. Sometimes nests early, even with snow lying.

LEFT: *MALE;* TOP: *FEMALE*

FACT FILE

LENGTH 16.5cm WINGSPAN 27–30.5cm WEIGHT 34–38g
FOOD Conifer seeds, especially spruce EGGS 4; off-white, sparsely spotted HABITAT Pine and spruce forests VOICE Call: explosive *chip chip*

JAN | FEB | MAR | APR | MAY | JUN | JUL | AUG | SEP | OCT | NOV | DEC

REED BUNTING

Emberiza schoeniclus

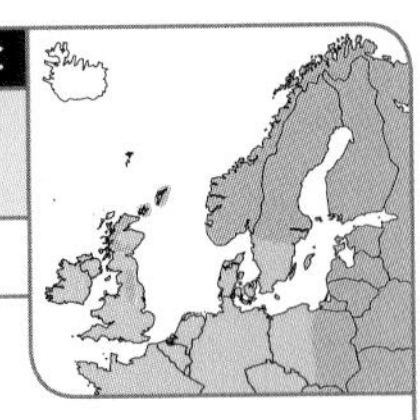

Occurs across region except Iceland. Mostly sedentary in British Isles. N and E birds migrate SW to winter in S and W Europe.

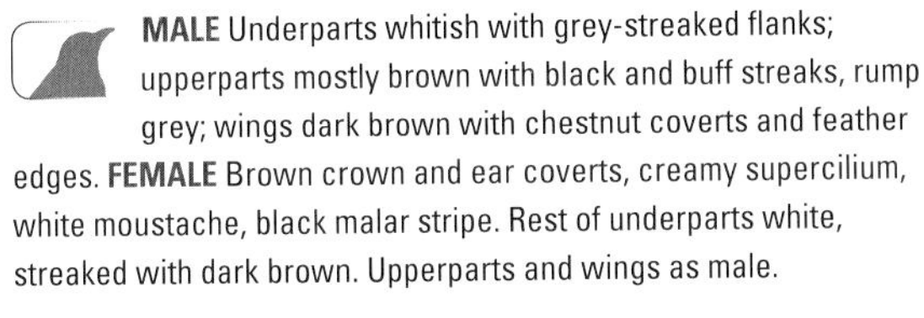

MALE Underparts whitish with grey-streaked flanks; upperparts mostly brown with black and buff streaks, rump grey; wings dark brown with chestnut coverts and feather edges. **FEMALE** Brown crown and ear coverts, creamy supercilium, white moustache, black malar stripe. Rest of underparts white, streaked with dark brown. Upperparts and wings as male.

Sociable in winter on drier ground. Sings from an exposed perch.

ABOVE: *FEMALE*
LEFT: *MALE*

FACT FILE

LENGTH 15–16.5cm
WINGSPAN 21–28cm
WEIGHT 16.6–25g
FOOD Seeds, invertebrates
EGGS 4–5, pale bluish with dark spots
HABITAT Thick cover by lakesides or riverbanks, willows by moorland streams
VOICE Call: high, soft *seeoo*. Song: repeated *tweek tweek tweek tititik*

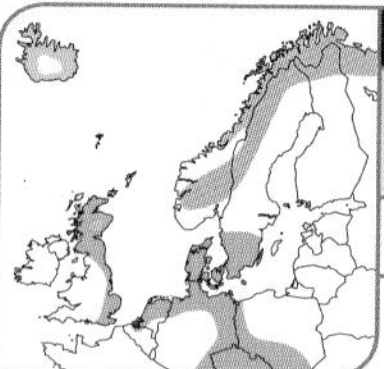

JAN | FEB | MAR | APR | MAY | JUN | JUL | AUG | SEP | OCT | NOV | DEC

SNOW BUNTING

Plectrophenax nivalis

Breeds in Scotland (<100 pairs), Iceland, Norway, far N Sweden, Finland, Russia and high-Arctic islands. Winters on Norwegian coast S from c. 65°N, around North Sea coasts, and more widely in Scotland.

SUMMER MALE White head, rump and underparts; white wings with triangular black wing-tip and bastard wing; black tail with white outer feathers; tawny tips to mantle. **WINTER MALE** Tawny colour extends over crown, ear coverts and sides of breast. **FEMALE** Patterned like winter male but browner, with less white on wings.

Always terrestrial. In winter, forms flocks for short migration south.

TOP: *JUVENILE*; ABOVE: *WINTER MALE*

FACT FILE

LENGTH 16–17cm WINGSPAN 32–38cm WEIGHT 33–50g FOOD Seeds EGGS 4–6; white, blotched brown HABITAT Breeds on treeless moors, tundra, bare mountains, N rocky coasts. Winters on rough land by coasts and shorelines VOICE Call: *tirrtirrir-rip*

JAN | FEB | MAR | APR | MAY | JUN | JUL | AUG | SEP | OCT | NOV | DEC

YELLOWHAMMER

Emberiza citrinella

MALE Yellow head, often almost unmarked; warm brown upperparts, streaked darker; chestnut rump shows well in flight; yellow underparts with chestnut on sides of breast and flanks; brown-black wings and tail, which has white outer feathers. **FEMALE** Duller, browner above and less yellow below; more heavily marked with olive on crown, ear coverts and malar stripe; underparts noticeably streaked, blackish rather than chestnut; rump reddish brown.

Winters on stubble, fallow land and stackyards.

Found almost throughout, but not high Arctic or Iceland. Mostly winters within breeding area; E and N birds tend to disperse further.

ABOVE: *FEMALE*
LEFT: *MALE*

FACT FILE

LENGTH 16–16.5cm WINGSPAN 23–29.5cm WEIGHT 23–35g FOOD Invertebrates, seeds EGGS 3–5, pale with dark scribbles HABITAT Farmland, heaths, coastal fields and scrub VOICE Call: ringing *zit*. Song: commonly rendered as 'little bit of bread and no cheese'

FURTHER INFORMATION

References and Further Reading

Aulén, G. (1996). *Where to Watch Birds in Scandinavia*. Hamlyn.

British Library (2009). *Coastal Birds: a Guide to Bird Sounds of the British Coast*. Audio CD and booklet. British Library.

British Library (2009). *Countryside Birds: a Guide to Bird Sounds of the British Countryside*. Audio CD and booklet. British Library.

Gooders, J. (1994). Where *to Watch Birds in Britain and Europe*. Christopher Helm.

Harrap, S. and Redman, N. (2003). *Where to Watch Birds – Britain*. Christopher Helm.

Holden, P. and Cleeves, T. (2010). *RSPB Handbook of British Birds* (3rd edition). RSPB.

Milne, P. and Hutchinson, C. (2009). *Where to Watch Birds – Ireland*. Christopher Helm.

Mullarney, K., Svensson, L., Zetterström, D. and Grant, P.J. (2010). *Collins Bird Guide* (2nd edition). HarperCollins.

Sterry, P. (2004). *Complete British Birds*. HarperCollins.

Toms, M. (2003). *The BTO/CJ Garden BirdWatch Book*. BTO.

Wheatley, N. (2000). *Where to Watch Birds in Europe and Russia*. Christopher Helm.

Many books that deal in detail with one family or a single species (monographs) are also available. Famous titles over the years have included the following:

Lack, D. (1946). *The Life of the Robin*. Witherby; reprinted in 1953 by Penguin.

Ratcliffe, D. (1980). *The Peregrine Falcon*. Poyser.

Summers-Smith, J.D. (1963). *The House Sparrow*. Collins.

Useful Addresses and Contacts

Each country has its own national birdwatching society, affiliated as a 'partner' to BirdLife International, the world's leading bird research and conservation organisation. In the United Kingdom, the BirdLife International partner and leading bird protection organisation is the Royal Society for the Protection of Birds. If you are interested in helping as a volunteer on a bird survey of populations, movements or ecology, contact the United Kingdom's expert in that field, the British Trust for Ornithology. The contact details of each of these organisations are given below:

BirdLife International
Wellbrook Court
Girton Road
Cambridge
CB3 0NA
www.birdlife.org

Royal Society for the Protection of Birds (RSPB)
The Lodge
Sandy
Bedfordshire
SG19 2DL
www.rspb.org.uk

British Trust for Ornithology (BTO)
The Nunnery
Thetford
Norfolk
IP24 2PU
www.bto.org

INDEX

Tick boxes are included next to the English name of each species so you can mark off species that you have seen.